Grade 1

Reveal MATH®

Student Practice Book

Mc
Graw
Hill

mheducation.com/prek-12

Send all inquiries to:
McGraw Hill
8787 Orion Place
Columbus, OH 43240

ISBN: 978-0-07-693704-2
MHID: 0-07-693704-6

Printed in the United States of America.

8 9 10 BRR 26 25 24 23 A

Grade 1
Table of Contents

Unit 4
Addition within 20: Facts and Strategies

Lessons

Unit 5
Subtraction within 20: Facts and Strategies

Lessons

Unit 6
Shapes and Solids
Lessons

Unit 7
Meanings of Addition
Lessons

Unit 8
Meanings of Subtraction
Lessons

Unit 9
Addition within 100
Lessons

Unit 10
Compare Using Addition and Subtraction
Lessons

Unit 11
Subtraction within 100

Lessons

Unit 12
Measurement and Data

Lessons

Unit 13
Equal Shares

Lessons

Additional Practice

Name _____

Review

You can find patterns when counting by 1s.

| 22 | 23 | 24 | 25 | 26 | 27 | 28 | **29** | **30** | 31 |

- The ones go up by 1 from 2 to 9. After 9, the ones start again at 0.

- The tens stay the same until the ones start again at 0. The tens go up by 1 each time the ones start again at 0.

Count. What pattern do you notice?

1.

| 17 | 18 | 19 | 20 | 21 | 22 | 23 | 24 | 25 | 26 |

2.

| 90 | 91 | 92 | 93 | 94 | 95 | 96 | 97 | 98 | 99 |

3. Emma is counting by Is from 36 to 45. Circle the numbers that Emma will say. Cross out any numbers that Emma will *not* say.

39   50   35   44   40   48

4. Count. What are the missing numbers? Write the missing numbers.

| 54 | | | 57 | 58 | 59 | | |

Explain the pattern.

5. Nigel is counting by Is. He starts at 79. He says the next number is 70. How do you respond to Nigel?

Math @ Home Activity

Give your child many opportunities to find patterns when counting by Is. Write a series of 2-digit numbers on self-sticking notes, leaving out two to three of the numbers. Ask your child to complete the counting pattern by writing the missing numbers on self-sticking notes and placing them in the correct order in the series. Repeat with another series of 2-digit numbers.

Lesson 2-2

Additional Practice

Name _____

Review

You can use a number chart and counting patterns to help you count.

Count by 1s. What 4 numbers come after 99?

1	2	3	4	5	6	7	8	9	10
11	12	13	14	15	16	17	18	19	20
21	22	23	24	25	26	27	28	29	30
31	32	33	34	35	36	37	38	39	40
41	42	43	44	45	46	47	48	49	50
51	52	53	54	55	56	57	58	59	60
61	62	63	64	65	66	67	68	69	70
71	72	73	74	75	76	77	78	79	80
81	82	83	84	85	86	87	88	89	90
91	92	93	94	95	96	97	98	99	100
101	102	103	104	105	106	107	108	109	110
111	112	113	114	115	116	117	118	119	120

100, 101, 102, and 103 come after 99.

---

1. Use counting patterns and the number chart. Start at 64. What are the next 4 numbers?

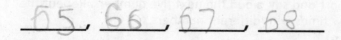

___65___, ___66___, ___67___, ___68___

Count by Is. What numbers come next? Use a number chart to help you.

2. 81, __82__, __93__, __84__, __85__

3. 106, __107__, __108__, __109__, __110__ ✓

4. 53, __54__, __55__, __56__, __57__

5. 92, __93__, __94__, __95__, __96__

✓6. Olga starts counting at ⟨109⟩. What are the next 3 numbers she counts?

   __110__, __111__, __112__

7. Explain the counting pattern that you notice after 100.

Additional Practice

Name _____

Review

You can use a number line to show counting patterns.

Start at 27. Count by 1s. Which numbers come next?

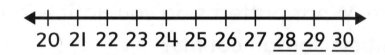

20 21 22 23 24 25 26 27 <u>28</u> <u>29</u> <u>30</u>

After 27, the numbers 28, 29, and 30 come next.

---

1. Start at 22. Count by 1s. Which numbers come next?

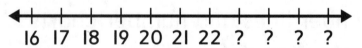

16 17 18 19 20 21 22 ? ? ? ?

A. 21, 22, 23, 24   **B.** 20, 21, 22, 23

C. 23, 24, 25, 26   **D.** 27, 28, 29, 30

2. Start at 77. Count by 1s. Which numbers come next?

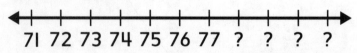

71 72 73 74 75 76 77 ? ? ? ?

A. 87, 97, 107, 117   **B.** 78, 79, 80, 81

C. 80, 82, 84, 86   **D.** 88, 99, 110, 111

3. Start at 102. Count by 1s. Which numbers come next?

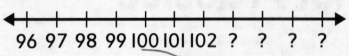

96 97 98 99 100 101 102 ? ? ? ?

A. 103, 104, 105, 106

B. 113, 114, 115, 116

C. 104, 106, 108, 110

D. 98, 99, 100, 101

4. Jordan counts 53 ducks. Then she counts 3 more. What patterns do you notice in the numbers Jordan counts?

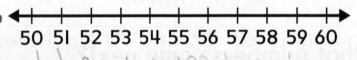

50 51 52 53 54 55 56 57 58 59 60

the number in the ones was ching, is

5. How are number lines and number charts alike? How are they different?

Lesson 2-4

Additional Practice

Name _____

Review

You can use counting patterns to help you read and write numbers to 120.

107, 108, __109__, __110__, __111__

- The ones go up by 1 to 9, then start again at 0.
- The tens stay the same. Then they go up by 1 when the ones start again at 0.

Count by 1s. What numbers come next?

1. 19, __20__, __21__, __22__, __23__

2. 39, __40__, __41__, __42__, __43__

3. a. 101, __102__, __103__, __104__, __105__

   b. Explain how you know.

4. a. 57, __58__, __59__, __60__, __61__

   b. Explain how you know.

Write the correct answer.

5. Count from 84 to 95 by 1s. What number comes next after (95)?

96

6. Count from 106 to (119) by 1s. What number comes next after 119?

120 ✓

7. Jewel is counting by 1s. She says 111 comes after 110. How do you respond to Jewel?

Yes

8. Think about counting patterns. How will the tens and ones change after 89?

Math @ Home Activity

Help your child find patterns when reading and writing numbers. First, give your child a number. Tell him or her to count by 1s and write the next three numbers. Then ask your child to identify any patterns he or she sees in the numbers. Repeat with different starting numbers.

Lesson 2-5
Additional Practice

Name _____

Review

You can count objects and write how many.

- Putting objects in a group can help you count them.

- These turtles are in groups of 10.

- Count the turtles.

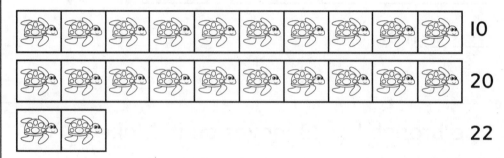

10

20

22

There are 22 turtles.

---

1. How many frogs are there?

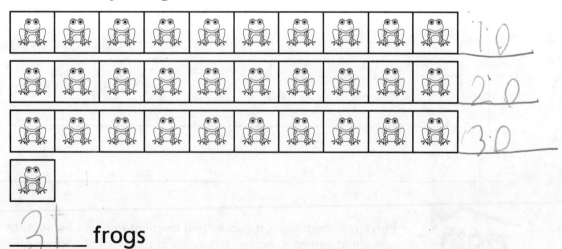

10

20

30

____31____ frogs

How many objects are there?

2.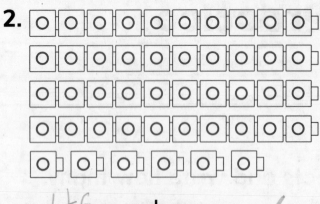

_46_ cubes ✓

3.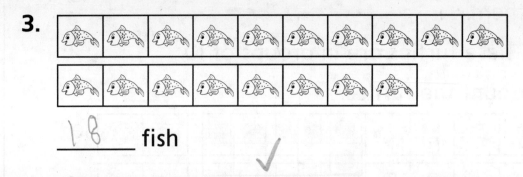

_18_ fish ✓

4. A tree branch has ⑬ leaves on it. Make a drawing to show how many leaves.

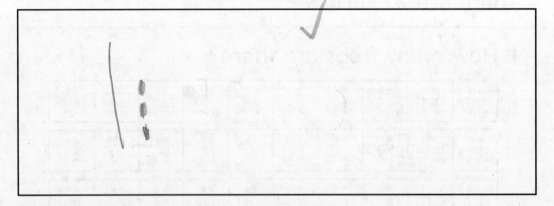

✓

Math @ Home Activity

Provide opportunities for your child to count groups of objects at home. For example, gather a handful of paper clips or dry beans. Have your child place the objects in rows of 10 and then count the total number of objects. Repeat with a different number of objects.

Additional Practice

Name _____

Review

You can use ten and some ones to make numbers 11 to 19. Numbers with one ten and some ones are *teen numbers*.

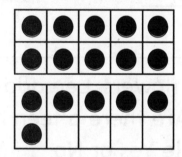

- The top ten-frame shows 1 group of ten.

- The bottom ten-frame shows 6 ones.

- 1 group of ten and 6 ones is 16.

How many counters?

1.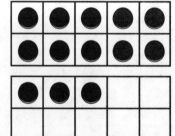

_____1_____ group of ten
and ___3___ ones
is __13__.

2.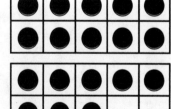

_____1_____ group of ten
and ___8___ ones
is __18__.

Draw counters on the ten-frames to show how many. Write the number.

3.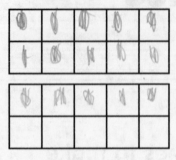

4.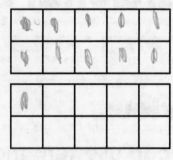

I group of ten and
5 ones is ___15___. ✓

I group of ten and
I one is ___11___. ✓

---

5. There are 10 flowers. Lee adds more flowers to make 12. Are there a teen number of flowers? Circle Yes or No.

(Yes)   No   ✓

6. There are 6 dogs at a park. A dog walker brings 3 more dogs. Now there are 9 dogs. Are there a teen number of dogs? Circle Yes or No. Explain your thinking.

Yes   (No)   ✓

Math @ Home Activity

Create a game for the teen numbers. On a set of cards, write the numbers 11–19. Cut the last two sections off each of two egg cartons, or draw two ten-frames on a sheet of paper. Have your child draw a card, and then have him or her show the teen number using dry beans or cereal. Repeat with other teen numbers.

Lesson 3-2

Additional Practice

Name ______________________

Review

You can make groups of ten.

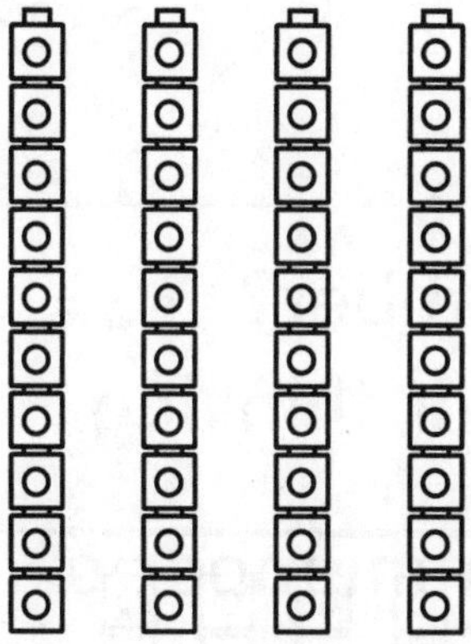

10 20 30 40

4 tens and 0 ones is 40.

How many tens and ones?

1.

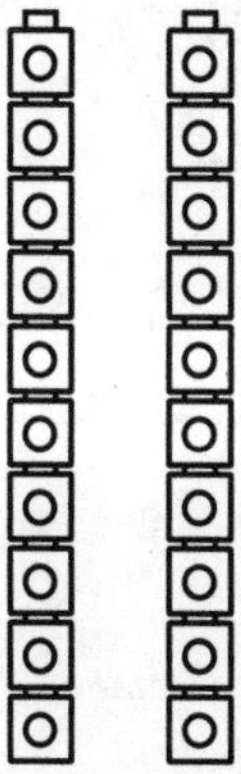

__2__ tens and

__0__ ones is

__20__.

2.

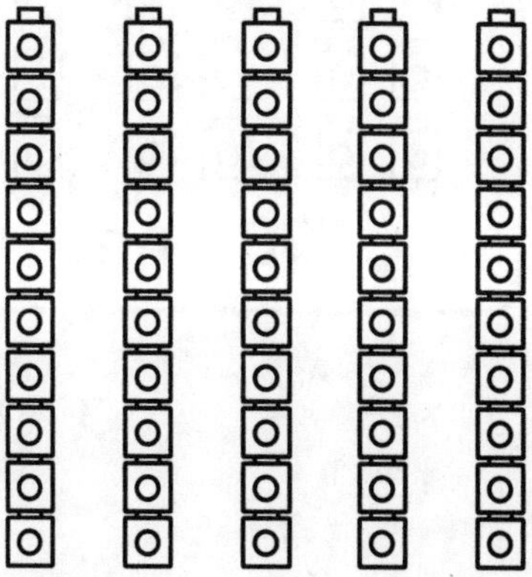

__5__ tens and

__0__ ones is

__50__.

3. Kira has 3 boxes. She puts 10 books in each box. How many books does she have?

_____30_____ books

4. A teacher puts students into groups of 10. If there are ⑦ groups of students, how many students are there?

_____70_____ students

5. Do these show the same number? Circle Yes or No. Explain your thinking.

30 ones

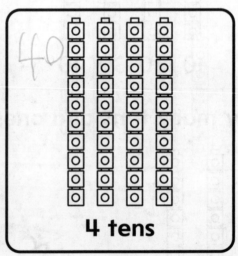

4 tens

Yes   No

Math @ Home Activity

Provide opportunities for your child to identify tens. Using string and beads or macaroni, work with your child to create 9 strands with 10 beads or pieces of macaroni on each strand. Then have your child place some strands in a row. Ask your child to count the number of tens and write the number they represent. Repeat the activity a few times with different numbers of strands.

Additional Practice

Name _____

Review

You can use cubes to show a 2-digit number as tens and ones.

Saul has 58 marbles. He can use cubes to show 58 with tens and ones.

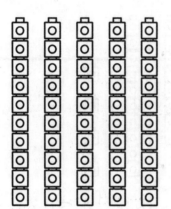

5 tens and 8 ones is 58.

How many? Circle the tens. Then write numbers to show how many.

1.

___4___ tens and ___4___ ones is ___44___.

How many? Circle the tens. Then write numbers to show how many.

2.

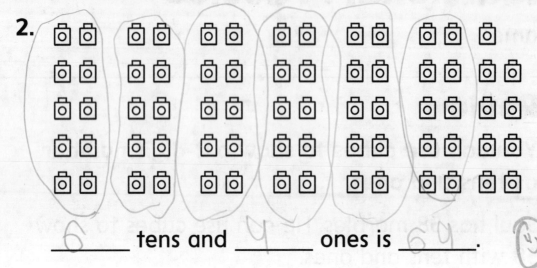

_____6_____ tens and _____4_____ ones is _____69_____.

3.

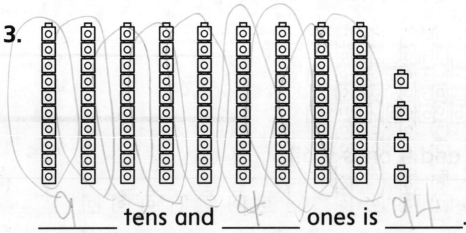

_____9_____ tens and _____4_____ ones is _____94_____.

4. A farmer sells bags of 10 apples. He has 8 full bags of apples and 7 apples left over. How many apples does he have?

_____87_____ apples

Using blank paper, draw and cut out single squares and strips of ten squares. Arrange a group of these tens and ones to show a number. Have your child point to each ten or one to count the number in the group. Then have your child write the number. Repeat the activity with different numbers.

Additional Practice

Name _____

Review

You can show 2-digit numbers with tens and ones.

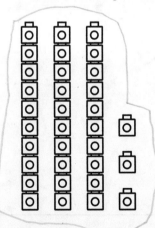

tens	ones
6	4

6 tens and 4 ones is 64.

---

1. Circle the cubes that show 33.
   How many tens? How many ones?

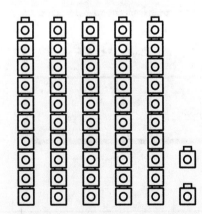

33 is ___3___ tens and ___3___ ones.

Use the number 86 to answer the questions.

2. What is the value of the 8?

_____8_____ tens or _____80_____

3. What is the value of the 6?

_____6_____ ones or _____6_____

4. Is 86 a 2-digit number? Circle Yes or No.

Yes   No

---

5. Ron uses cubes to show a number. How can you write his number in the place-value chart?

tens	ones
9	7

Math @ Home Activity

Work with your child to help him or her explain how to show 2-digit numbers with tens and ones. Write a 2-digit number at the top of a sheet of paper. Have your child draw cubes to show the number and complete this sentence: _____ is _____ tens and _____ ones.

Additional Practice

Name _____

Review

You can show the tens and ones in a number in different ways.

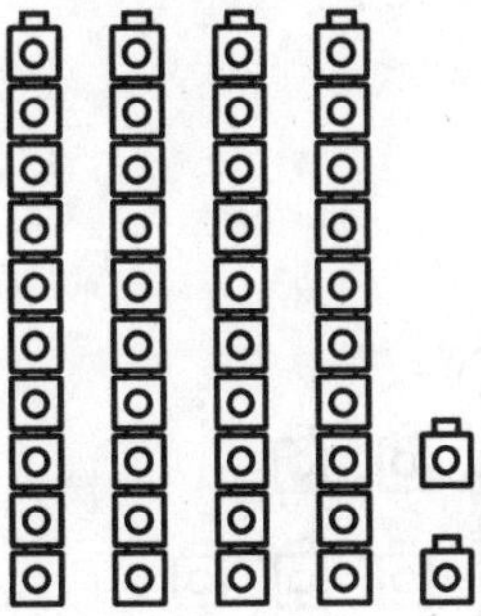

42 = 4 tens and 2 ones 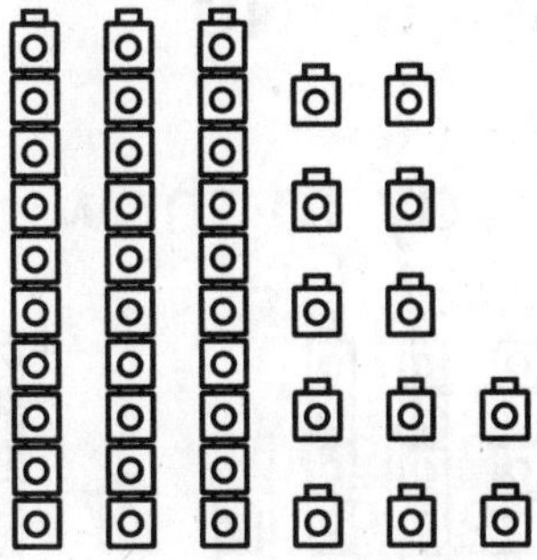 | 42 = 3 tens and 12 ones 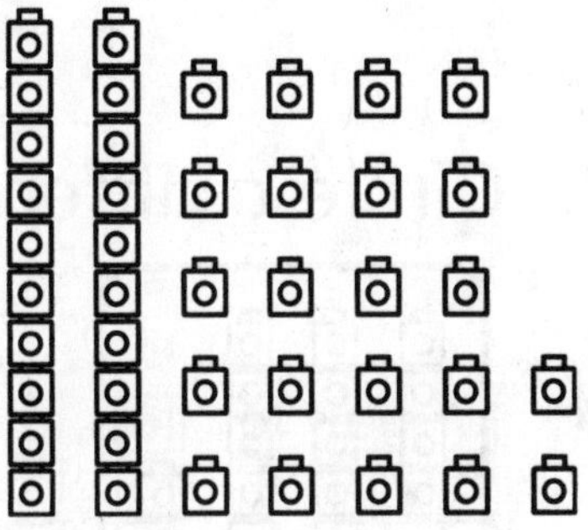 | 42 = 2 tens and 22 ones

---

1. How can you show 57 in different ways? Use connecting cubes to help.

   0 tens and __57__ ones

   I ten and __47__ ones

   2 tens and __37__ ones

   3 tens and __27__ ones

   4 tens and __17__ ones

   5 tens and __7__ ones

Wow.
☺

2. Do these show the same number?
Circle Yes or No.

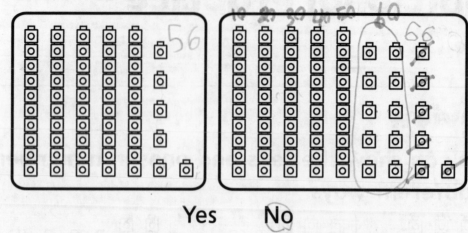

Yes   No

3. Circle a different way to show 34.

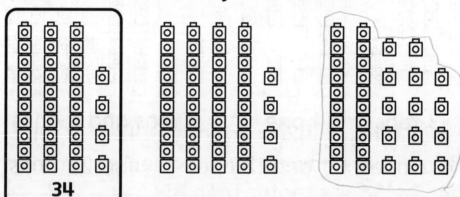

4. There are 68 students in the gym. What are two ways you can show the number of students?

_____6_____ tens and _____8_____ ones is _____68_____.

_____5_____ tens and _____10_____ ones is _____68_____.

Math @ Home Activity

With your child, draw and cut out single cubes and strips of ten cubes from paper. Display a group of tens and some ones. Ask your child to identify the number shown. Then encourage them to represent the same number using different tens and ones.

Additional Practice

Name
_____

Review

You can compare two numbers to determine which is greater.

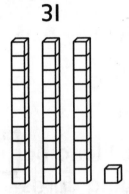

31

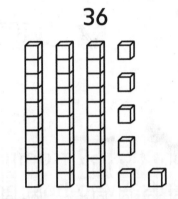

36

3 tens and 1 one   3 tens and 6 ones

Both numbers have 3 tens. 6 ones is greater than 1 one. So, 36 is greater than 31.

---

1. Circle *is greater than, is less than,* or *is equal to.*

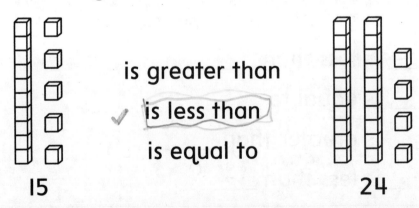

is greater than

✓ is less than

is equal to

15   24

2. Henyer writes the numbers 76 and 78 on a sheet of paper. Which number is less?

✓ 76

3. Circle the phrase to complete the comparison, 63 _____ 61.

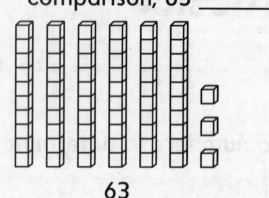 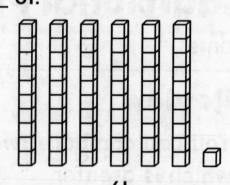

63   61

✓ (is greater than)

is less than

is equal to

4. Sam has 59 trading cards. Elena has 61 trading cards. Who has more trading cards?

Sam   (Elena)

_____

Write numbers to make each sentence true.

✓ **5.** __69__ is greater than __20__.

✓ **6.** __4__ is less than __100__.

✓ **7.** __70__ is equal to __70__.

✓ **8.** __11__ is greater than __2__.

✓ **9.** __10__ is less than __99__.

_____

Math @ Home Activity

Have your child roll a number cube four times to create two 2-digit numbers. Have him or her write the numbers on a sheet of paper and determine the relationship between them using the words *is greater than*, *is less than*, or *is equal to*. Repeat the activity several times.

Additional Practice

Name _____

Review

You can compare numbers on a number line.

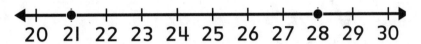

20 21 22 23 24 25 26 27 28 29 30

- On a number line, the number to the right is greater.

- The number to the left is less.

- 28 is greater than 21.

Draw dots on the number line to compare the numbers. Write *is greater than, is less than,* or *is equal to.*

1.

25 26 27 28 29 30 31 32 33 34 35 36 37 38 39 40

29 ___is less___ 38.

2.

55 56 57 58 59 60 61 62 63 64 65 66 67 68 69 70

70 ___is great___ 58.

Use the number line to compare the numbers.
Circle the number that is less.

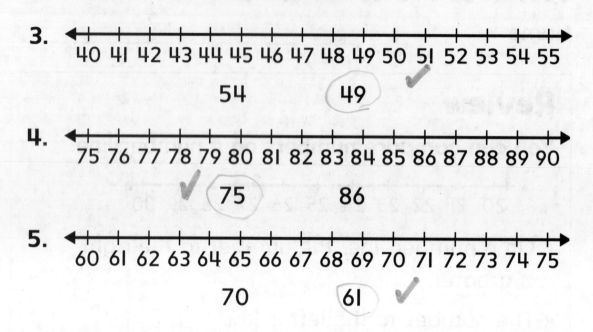

3.

40 41 42 43 44 45 46 47 48 49 50 51 52 53 54 55

54   (49) ✓

4.

75 76 77 78 79 80 81 82 83 84 85 86 87 88 89 90

✓ (75)   86

5.

60 61 62 63 64 65 66 67 68 69 70 71 72 73 74 75

70   (61) ✓

6. Anika has more than 66 trading cards.
She has less than 71 trading cards.

Use the number line to show how many
trading cards Anika might have.

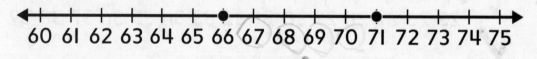

60 61 62 63 64 65 66 67 68 69 70 71 72 73 74 75

✓

Math
@ Home
Activity

On an erasable surface, create a number line that spans 20 numbers.
Place dots on the line over two numbers. Ask your child to identify
which number is greater. Repeat the activity, this time with your child
placing the dots and you determining the greater number.

Additional Practice

Name _____

Review

You can use > (is greater than),
< (is less than), and = (is equal to)
to compare numbers.

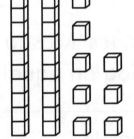

28 is less than 41

28 < 41

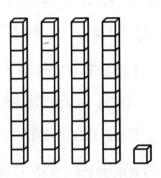

Circle the correct symbol.

1. is equal to   >   <   (=)

2. is greater than   (>)   <   =   ☺

3. is less than   >   (<)   =

Compare the numbers. Write >, <, or =.

4. 98 ⊘ 98

5. 26 ⊘ 19

6. 50 ⊘ 70

7. 11 ⊘ 11

Compare the numbers. Write >, <, or =.

8. 21 ◯ 39

9. 38 ◯ 31

10. 53 ◯ 53

11. 64 ◯ 54

12. 91 ◯> 77

13. 42 ◯ 43

14. 30 ◯ 30

15. 86 ◯> 76

Circle the correct answer. Write >, <, or = to compare the numbers.

16. Matt jogs 47 miles. Chen jogs 51 miles. Masha jogs 47 miles. Which two people jog an equal number of miles?

    Matt   Chen   Masha

    47 ◯= 47

17. There are 68 red balls and 71 green balls in a gym. Are there more red balls or green balls?

    green balls   red balls

    68 ◯< 71

Math @ Home Activity

Look for situations around your home where your child can practice using the symbols >, <, and =. For example, if you have 10 oranges and 11 apples, ask your child to compare the numbers. Your child can carry around self-sticking notes and a pencil to draw the correct comparison symbols for different situations.

Additional Practice

Name _____

Review

You can count or add to get a total.

There are 2 pigs in a puddle. There are 3 pigs in another puddle. How many pigs are there?

Count to get the total.

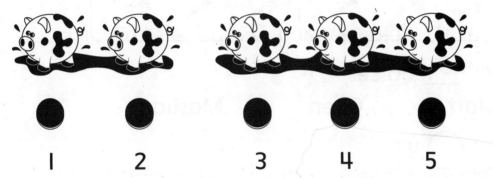

Add to get the total.

$2 + 3 = 5$

There are 5 pigs.

How many? Write the number.

I.

2.

_____ sheep   _____ birds

3. How many cows? Write numbers to match the picture.

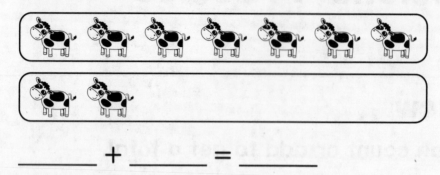

_____ + _____ = _____

4. How many horses? Write the number.

_____ horses

5. Draw a picture to show how to add 6 + 5. Write numbers to match the picture you drew.

_____ + _____ = _____

Math @ Home Activity

Provide opportunities for your child to add at home. For example, during mealtime, have your child count the number of plates or add the number of spoons and forks on the table.

Additional Practice

Name _____

Review

You can use a number line to count on.

Kaci biked 7 laps. Then she biked 2 more laps. How many total laps did she bike?

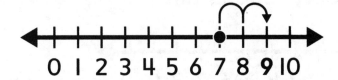

$7 + 2 = 9$

Kaci biked 9 laps total.

What is the sum? Use the number line to add.

1. $1 + 8 =$ _____

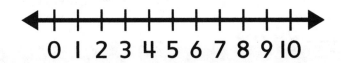

2. $7 + 5 =$ _____

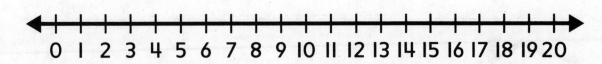

What is the sum? Use the number line to add.

3. $6 + 1 =$ _____

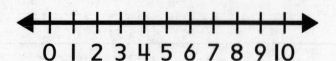

0 1 2 3 4 5 6 7 8 9 10

4. Cory reads 7 pages of his book in the morning and 3 pages in the evening. How many pages does Cory read in all?

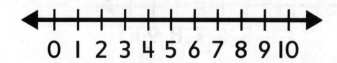

0 1 2 3 4 5 6 7 8 9 10

$7 + 3 =$ _____ pages

5. Donya scores 8 points in one game. She scores 6 points in another game. How many points does she score in all?

$8 + 6 =$ _____ points

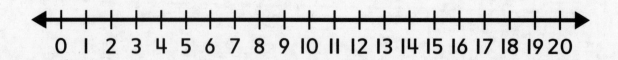

0 1 2 3 4 5 6 7 8 9 10 11 12 13 14 15 16 17 18 19 20

Math @ Home Activity

Provide opportunities for your child to use a number line to add at home. For example, have your child add numbers with a sum up to 20 on a number line using a small object, such as a bean, as a marker.

Additional Practice

Name _____

Review

You can use doubles to add.

Mo makes 4 fruit cups. Seth makes 4 fruit cups. How many fruit cups do they make in all?

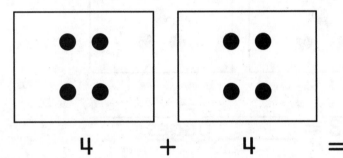

$$4 \quad + \quad 4 \quad = 8$$

Mo and Seth make 8 fruit cups.

What is the sum?

1. $7 + 7 =$ _____

2. $2 + 2 =$ _____

3. $6 + 6 =$ _____

4. $5 + 5 =$ _____

5. $8 + 8 =$ _____

6. $1 + 1 =$ _____

7. $9 + 9 =$ _____

8. $3 + 3 =$ _____

How many dots are there? Write numbers to match the picture.

9.

_____ + _____ = _____

10.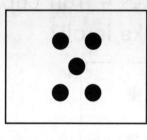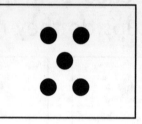

_____ + _____ = _____

11. Temple throws 9 baseballs. Rashad throws 9 baseballs. How many baseballs do they throw in all?

_____ + _____ = _____ baseballs

12. Elvin buys 8 books. Alicia buys 8 books. How many books do they buy in all?

_____ + _____ = _____ books

Math @ Home Activity

Ask your child to find a sum up to 18 in situations in which both addends are the same. For example, if you and your child both have 6 carrots, ask him or her to determine how many carrots you have in all.

Additional Practice

Name _____

Review

You can use doubles to help you add near doubles.

$$5 + 6 = ?$$

Use a doubles fact that you already know.

- Add the double: $5 + 5 = 10$
- Then add 1 more: $5 + 6 = 11$

How can you use doubles to add? Write the sum.

1. $2 + 1 = $ _____

2. $5 + 7 = $ _____

3. $4 + 5 = $ _____

4. $7 + 8 = $ _____

5. $5 + 3 = $ _____

6. $6 + 7 = $ _____

What is the sum?

7. $9 + 8 =$ _____

8. $8 + 6 =$ _____

9. $6 + 4 =$ _____

10. $9 + 7 =$ _____

---

11. Lyla eats 3 carrots. Then she eats 2 more carrots. How many carrots does Lyla eat in all?

_____ + _____ = _____ carrots

12. Raven has 8 tickets. Then she earns 6 more tickets. How many tickets does Raven have in all?

_____ + _____ = _____ tickets

Math @ Home Activity

Write an addition equation with the sum missing. Ask your child to identify a doubles fact he or she can use to help find the sum. After solving the equation, have your child check his or her answer by using small objects, such as paper clips, to represent the equation. Repeat with different addition equations with sums up to 17.

Additional Practice

Name _____

Review

You can make a 10 to add.

Carl has 8 black magnets and 4 white magnets.
How many magnets does he have in all?

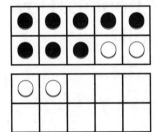 

$$10 + 2 = 12$$

Carl has 12 magnets in all.

What is the sum? Make a 10 to add.

1. $6 + 5 =$ _____

2. $9 + 6 =$ _____

What is the sum? Show how to make a 10 to add.

3. 7 + 6 = _____

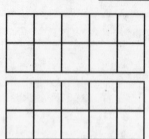

4. 4 + 9 = _____

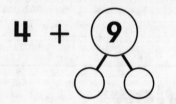

Complete the number bond to make a 10 to add.

5. Kelly has 8 pencils. Cari has 7 pencils. How many pencils do they have in all?

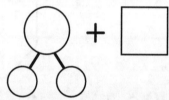

How can you make a 10 to add?

Kelly and Cari have _____ pencils in all.

Math @ Home Activity

With your child, practice adding numbers using ten-frames. Using two muffin tins with 10 openings each, two egg cartons with two sections removed each, or two ten-frames you draw, have your child make 10 to add numbers with sums from 11 to 18.

Additional Practice

Name _____

Review

You can add two numbers in different ways.

Aiko sees 7 ants and 6 bees in her yard. How many insects does she see in all?

You can count on using a number line.

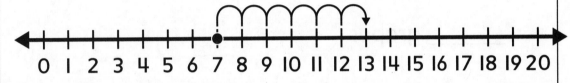

You can make a 10.

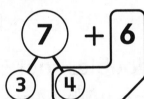

Aiko sees 13 insects in all.

What is the sum? Tell or show how you added.

1. $2 + 9 =$ _____

2. $8 + 5 =$ _____

What is the sum? Tell or show how you added.

3. Riley sleeps for 9 hours on Monday night. She sleeps for 8 hours on Tuesday night. How many hours does she sleep both nights in all?

_____ hours

4. Sima eats 6 green grapes and 6 purple grapes. How many grapes does she eat?

_____ grapes

5. Maggie uses the doubles fact 5 + 5 to add 2 + 3. Do you agree with Maggie's work? Show or explain your thinking.

Math @ Home Activity

Practice using different addition strategies at home. For example, ask your child, "If you have 5 red blocks and 6 yellow blocks, how many blocks do you have in all?" Have your child use a number line or near doubles to add. Then have your child choose a different strategy to solve the problem, such as make a 10.

Additional Practice

Name _____

Review

You can add two addends in any order.

$3 + 2 = 5$ $2 + 3 = 5$

You can add 3 and 2 in any order. The sum is 5.

What is the sum?

1. $8 + 1 =$ _____   **2.** $5 + 7 =$ _____

$1 + 8 =$ _____ $7 + 5 =$ _____

3. $4 + 6 =$ _____   **4.** $9 + 8 =$ _____

$6 + 4 =$ _____ $8 + 9 =$ _____

---

5. Which has the same sum as $7 + 2$?

A. $2 + 6$ **B.** $3 + 7$

C. $2 + 7$ **D.** $1 + 9$

6. Which has the same sum as 3 + 9?

   A. 8 + 2   **B.** 9 + 3

   C. 7 + 6   **D.** 4 + 9

7. What is another way to add 6 + 7 that has the same sum?

   _____ + _____ = _____

8. One fish tank has 7 fish. Another fish tank has 8 fish. How many fish are in the tanks? Write numbers to show two ways to add.

   _____ + _____ = _____

   _____ + _____ = _____

9. Mr. Thompson reads for 9 minutes in the morning. Then he reads for 5 minutes. How long did he read? Write numbers to show two ways to add.

   _____ + _____ = _____

   _____ + _____ = _____

Math @ Home Activity

Have your child demonstrate that the order in which numbers are added does not matter. For example, have your child explain why adding 3 red apples and 8 yellow apples results in the same sum as adding 8 yellow apples and 3 red apples.

Additional Practice

Name _____

Review

You can add three addends in any order.

$1 + 3 + 7 = 11$ $3 + 1 + 7 = 11$ $7 + 1 + 3 = 11$

$1 + 7 + 3 = 11$ $3 + 7 + 1 = 11$ $7 + 3 + 1 = 11$

Add 1, 3, and 7 in any order. The sum is always 11.

What is the sum?

1.

   $5 + 1 + 5 =$ _____

2.

   $4 + 2 + 4 =$ _____

What is the sum?

3. $3 + 5 + 2 =$ _____   **4.** $7 + 3 + 8 =$ _____

5. $9 + 4 + 1 =$ _____   **6.** $7 + 5 + 5 =$ _____

---

7. Amy, Bonnie, and Tan buy apples at the store.
Find the total number of apples they buy.
Write the sum.

Amy's Apples	Bonnie's Apples	Tan's Apples
🍎 🍎 🍎 🍎	🍎 🍎 🍎 🍎 🍎 🍎	🍎 🍎 🍎 🍎 🍎

_____ apples

Additional Practice

Name _____

Review

You can use one addend and the sum to find an unknown addend.

There are 8 pennies in a jar. Wu adds more pennies to the jar. Now there are 15 pennies. How many pennies did Wu add?

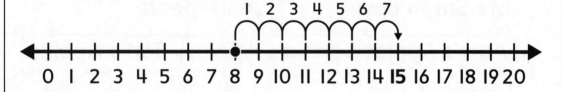

Start at 8. End at 15. $8 + ? = 15$

The unknown addend is 7. $8 + 7 = 15$

Wu added 7 pennies.

What is the unknown addend?

1. $3 + $ _____ $= 10$ **2.** $4 + $ _____ $= 12$

3. _____ $+ 6 = 11$ **4.** _____ $+ 7 = 10$

5. _____ $+ 5 = 9$ **6.** $2 + $ _____ $= 10$

What is the unknown addend?

7. $4 + \underline{\hspace{2cm}} = 13$ 8. $\underline{\hspace{2cm}} + 9 = 16$

---

9. Ms. Smith has 5 pears. She buys some more pears. Now she has 12 pears. How many more pears did she buy?

   Which matches the problem?

   $5 + ? = 12$ $5 + 12 = ?$

   What is the unknown addend?

   Ms. Smith buys $\underline{\hspace{2cm}}$ more pears.

10. There are 8 children on a ride. More children get on the ride. Now there are 16 children on the ride. How many more children got on the ride?

    Fill in the equation to show the problem. Then fill in the equation with the unknown addend.

    $\underline{\hspace{2cm}} + ? = \underline{\hspace{2cm}}$

    $\underline{\hspace{2cm}} + \underline{\hspace{2cm}} = \underline{\hspace{2cm}}$

Copyright © McGraw-Hill Education

Additional Practice

Name _____

> ## Review
>
> **You can use an equal sign (=) to show that the amounts on both sides are equal.**
>
> Jamal has 4 black feathers and 5 white feathers. Frida has 6 black feathers and 3 white feathers.
>
> Jamal's Feathers   Frida's Feathers
>
>  $4 + 5 = 6 + 3$
>
> $9 = 9$
>
> Jamal and Frida each have 9 feathers.

Do Olivia and Armand have an equal amount of crayons? Write each sum.

1. Olivia has 6 white crayons and 7 black crayons.

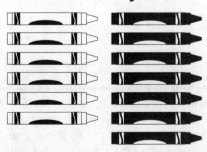

$6 + 7 =$ _____

2. Armand has 8 white crayons and 5 black crayons.

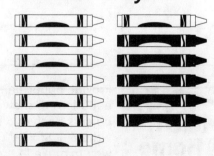

$8 + 5 =$ _____

3. Gwen has 6 gray rocks and 9 white rocks. James has 8 gray rocks and 7 white rocks. How can you show if they have the same number of rocks?

4. Em eats 7 green grapes and 5 red grapes. Jet eats 3 green grapes and 8 red grapes. Do they eat an equal number of grapes? Circle Yes or No. Draw a picture to show your thinking.

Yes   No

5. Milo says that 6 + 8 = 7 + 7. Draw a picture to show the equation is true.

Math @ Home Activity

Provide opportunities for your child to practice writing equal signs in equations to show equal amounts. For example, represent 6 + 8 using objects, such as 6 red crayons and 8 blue crayons. Have your child write 6 + 8 = 14. Then have them use the objects to show another way to make 14 and write the corresponding equation, for example, 9 + 5 = 14. Finally, help your child write an equation to show the two amounts are equal: 6 + 8 = 9 + 5.

Additional Practice

Name _____

Review

You can compare amounts to show if an equation is true or false.

Is this true or false? $4 + 7 \overset{?}{=} 7 + 4$

$4 + 7 = 11$

$7 + 4 = 11$

An equation is true when the amounts on both sides are equal, so $4 + 7 \overset{?}{=} 7 + 4$ is true.

Is the equation true or false? Circle the answer.

1. $1 + 9 \overset{?}{=} 6 + 3$

      True   False

2. $5 + 3 \overset{?}{=} 1 + 7$

      True   False

3. $8 + 7 \overset{?}{=} 6 + 9$

      True   False

4. $3 + 9 \overset{?}{=} 8 + 5$

      True   False

Is the equation true? Write Yes or No.

5. $9 + 4 \overset{?}{=} 8 + 4$ _____

6. $5 + 5 \overset{?}{=} 3 + 6$ _____

7. $9 + 7 \overset{?}{=} 8 + 8$ _____

8. Write 4 different addition expressions that make $6 + 2 =$ _____ + _____ true.

_____ + _____

_____ + _____

_____ + _____

_____ + _____

9. Aiden says that $7 + 6 \overset{?}{=} 4 + 9$ is true. Do you agree? Explain your thinking.

Math @ Home Activity

Have your child demonstrate understanding of equal amounts in equations. Provide him or her with small objects, such as uncooked beans, and an equal sign and two plus signs written on separate index cards. Encourage your child to show two equal addition expressions, such as $3 + 2 = 4 + 1$, using the small objects and cards.

Additional Practice

Name _____

Review

You can count or subtract to get a difference.

There are some penguins on ice. Some dive off the ice. How many penguins are still on the ice?

• Take away counters to find the difference.

$$5 - 2 = 3$$

There are 3 penguins left.

What numbers match the picture?
Complete the equation.

1. $4 - \underline{} = \underline{}$

2. $\underline{} - \underline{} = 4$

How many are left? Complete the equation.

3. There are some foxes. Some run away.
How many foxes are left?

10 − _____ = _____ foxes

4. There are some turtles. Some swim away.
How many turtles are left?

8 − _____ = _____ turtles

5. Draw a picture to show how to subtract 11 − 4.
What numbers match your picture? Complete
the equation.

_____ − _____ = _____

Math @ Home Activity

Provide opportunities for your child to use objects to subtract at home. For example, gather at least 12 marbles or other small items. Take away some of the marbles, and have your child write a subtraction equation to show the difference.

Additional Practice

Name _____

Review

You can count back to subtract.

• Use a number line to subtract 10 − 4.

• Start at 10.

• Count back by 4.

• End on the difference, 6.

$$10 - 4 = 6$$

5 6 7 8 9 10 11 12 13 14 15 16 17 18 19 20

How can you count back to subtract?
Use the number line to find the difference.

1. 8 − 2 = _____

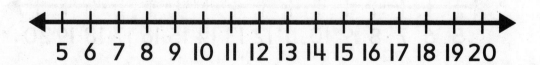

5 6 7 8 9 10 11 12 13 14 15 16 17 18 19 20

2. 13 − 6 = _____

5 6 7 8 9 10 11 12 13 14 15 16 17 18 19 20

How can you count back to subtract?
Use the number line to find the difference.

3. $11 - 3 =$ _____

4. $16 - 7 =$ _____

What is the difference? Use the number line to help.

5. $15 - 8 =$ _____

6. $14 - 2 =$ _____

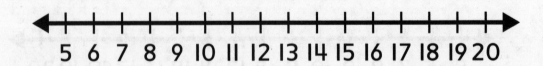

 **Math @ Home Activity**

Ask your child to explain how to use a number line to subtract. Using sidewalk chalk or masking tape, create a large number line. Have your child act out subtraction problems by moving back spaces to show the subtraction.

Additional Practice

Name _____

Review

You can count on to subtract.

• Use a number line to subtract $14 - 6$.

• Start at 6.

• Count on to the total, 14.

• The number of jumps is the difference.

$$14 - 6 = \mathbf{8}$$

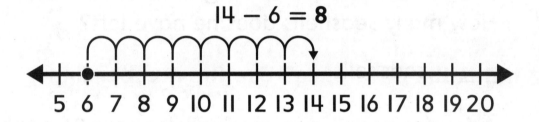

How can you count on to subtract?
Use the number line to find the difference.

1. $11 - 5 =$ _____

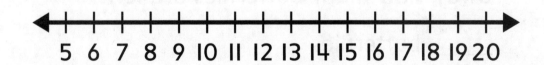

2. $15 - 8 =$ _____

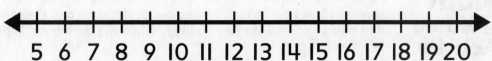

What is the difference? Use the number line to help.

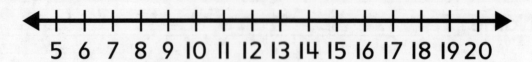

5 6 7 8 9 10 11 12 13 14 15 16 17 18 19 20

3. 18 − 8 = _____

4. 14 − 9 = _____

5. Malik has 12 seashells. He gives 5 to his sister. How many seashells does he have left?

_____ seashells

6. A farmer has 17 watermelons. He sells 8 at the market. How many watermelons are left?

_____ watermelons

7. There are 14 butterflies in a garden, and 6 fly away. How many butterflies are left?

_____ butterflies

Math @ Home Activity

Provide opportunities for your child to use a number line to count on to subtract. Draw a number line from 0 to 20. Choose a starting number. Have your child roll a number or dot cube and count on that many from the starting number to find the difference. Repeat as time allows.

Additional Practice

Name _____

Review

You can make a 10 to help you subtract.

Clark has 12 apples. He gives 3 apples to his mom. How many apples does Clark have left?

- Subtract 12 − 3 by making a 10.

- Break apart 3 into 2 and 1.

- Make a 10; 12 − 2 = 10

- 10 − 1 = 9

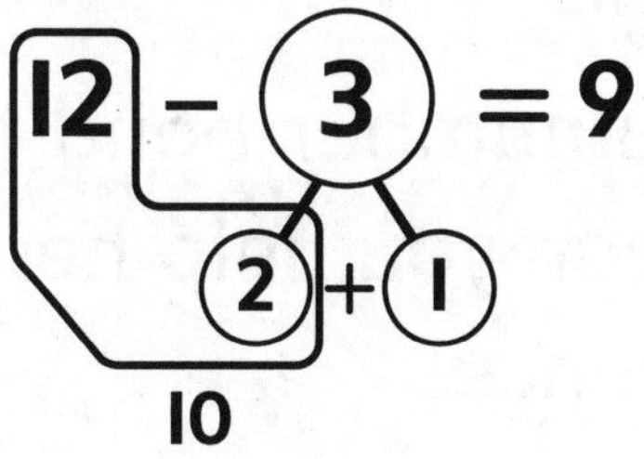

$$12 - \boxed{3} = 9$$

$$\boxed{2} + \boxed{1}$$

10

Clark has 9 apples left.

What is the difference?

1. 11 − 6 = _____

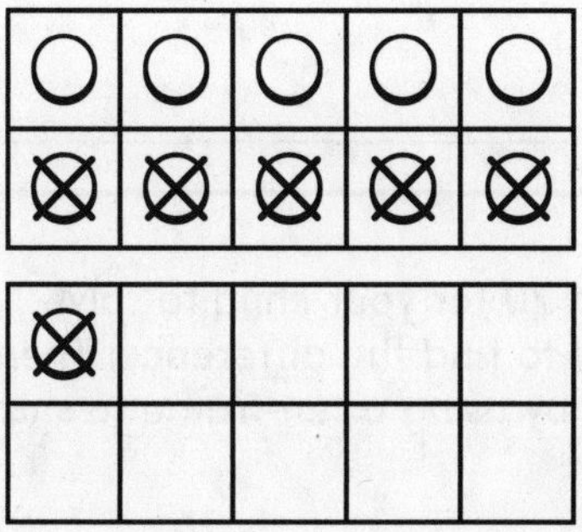

2. 13 − 7 = _____

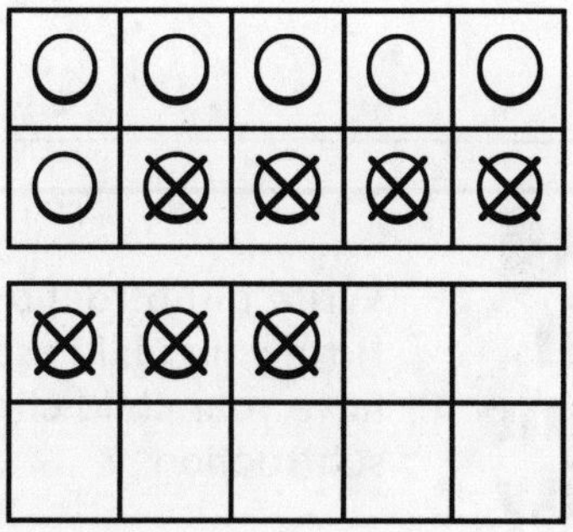

How can you make a 10 to subtract?
Show your thinking. Write the difference.

3. $16 - 8 =$ _____

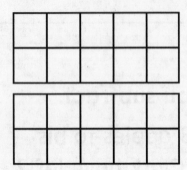

4. $14 - 5 =$ _____

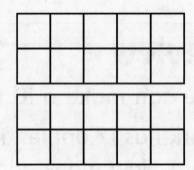

5. $18 - 9 =$ _____

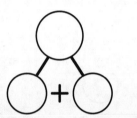

6. $15 - 7 =$ _____

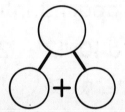

7. Renata uses a number bond to subtract
17 − 8. How can you help her find the
correct difference?

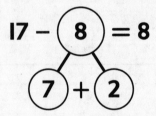

$17 - (8) = 8$

$(7) + (2)$

Math @ Home Activity

Write a subtraction problem within 20 for your child to solve.
Have your child use a number bond to find the difference. Then
have your child check their answer by using a ten-frame to show the
subtraction.

Additional Practice

Name _____

Review

You can use doubles and near doubles to solve subtraction equations.

There are 12 children sitting at a table. Then 5 children walk away. How many children are left at the table?

$$12 - 5 = ?$$

○ ○ ○ ○ ○ ○

○ │○ ○ ○ ○ ○│

- You know $6 + 6 = 12$. So, $12 - 6 = 6$.

- You take away 1 less in $12 - 5$ than in $12 - 6$.

- The difference of $12 - 5$ is 1 more than 6.
  So, $12 - 5 = 7$.

What is the difference? Write a double to help you subtract.

1. $6 - 3 =$ _____

2. $16 - 8 =$ _____

3. $18 - 9 =$ _____

4. $14 - 7 =$ _____

What is the difference? Explain how you used a double to help you subtract.

5. $10 - 6 = \underline{}$

6. $16 - 7 = \underline{}$

7. Zola reads 13 pages in her book for homework. She reads 7 pages before dinner. How many pages does Zola read after dinner?

    $\underline{}$ pages

Math @ Home Activity

Have your child roll a number or dot cube and use the number rolled to write a doubles fact. Then have your child write a related subtraction equation. For example, if a 5 is rolled, your child can write $5 + 5 = 10$ and $10 = 5 - 5$.

Additional Practice

Name _____

Review

You can use addition to help you subtract.

$$14 - 9 = ?$$

Make a related addition equation with an unknown addend.

$$9 + ? = 14$$

Find the unknown addend.

$$5 \quad 6 \quad 7 \quad 8 \quad 9 \quad 10 \quad 11 \quad 12 \quad 13 \quad 14 \quad 15 \quad 16 \quad 17 \quad 18 \quad 19 \quad 20$$

$9 + 5 = 14$, so $14 - 9 = 5$.

How can you use the addition equation to help you subtract? Write the difference.

1. $7 + 3 = 10$, so

$10 - 7 =$ _____.

2. $8 + 8 = 16$, so

$16 - 8 =$ _____.

3. $6 + 9 = 15$, so

$15 - 6 =$ _____.

4. $4 + 7 = 11$, so

$11 - 4 =$ _____.

What addition equation can help you subtract?
Write the addition equation. Write the difference.

5. $18 - 9 =$ _____

6. $9 - 4 =$ _____

7. $13 - 6 =$ _____

8. $17 - 8 =$ _____

9. There are 10 frogs on a log. 5 frogs jump into the pond. How many frogs are on the log now?

_____ frogs

10. Explain how you can add to help you subtract $14 - 6$.

Math @ Home Activity

Write subtraction problems within 20 on index cards and place them facedown on a table. With your child, take turns drawing index cards and finding each difference. On the index card, the solver should write an addition equation that can be used to solve the subtraction problem. If the problem is correctly solved, the solver keeps the card. If not, it is returned facedown to the table.

Additional Practice

Name _____

Review

You can use a fact triangle to make related facts.

The facts in a fact family all use the same three numbers.

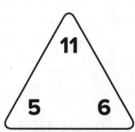

$5 + 6 = 11$

$6 + 5 = 11$

$11 - 5 = 6$

$11 - 6 = 5$

Complete the related facts for the fact triangle.

1.

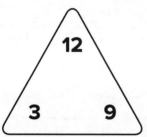

$3 + \underline{} = 12$

$9 + 3 = \underline{}$

$12 - \underline{} = 9$

$\underline{} - 9 = 3$

2.

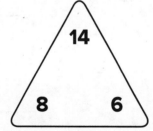

$8 + \underline{} = 14$

$6 + 8 = \underline{}$

$14 - \underline{} = 6$

$\underline{} - 6 = 8$

What is the fact family for the fact triangle? Write the facts.

3.

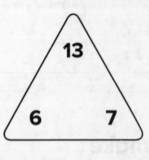

____ + ____ = ____

____ + ____ = ____

____ − ____ = ____

____ − ____ = ____

4.

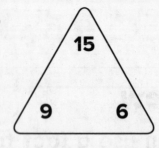

____ + ____ = ____

____ + ____ = ____

____ − ____ = ____

____ − ____ = ____

5. Write your own fact triangle and fact family.

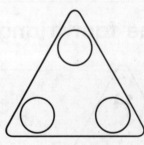

____ + ____ = ____

____ + ____ = ____

____ − ____ = ____

____ − ____ = ____

Additional Practice

Name _____

Review

You can find an unknown number in a subtraction equation in different ways.

$$16 - ? = 7$$

One way is to use addition to solve.

$$7 + ? = 16$$

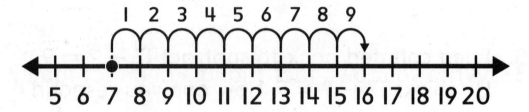

$7 + 9 = 16$, so $16 - 9 = 7$.

How can you complete the equation?
Tell or show how you solved.

1. $12 - 5 =$ _____

2. $14 -$ _____ $= 6$

How can you complete the equation?
Tell or show how you solved.

3. _____ = 15 − 7

4. 11 − _____ = 5

5. Luca counted back to subtract 17 − ? = 9.
He thinks 17 − 9 = 9. How can you respond
to Luca?

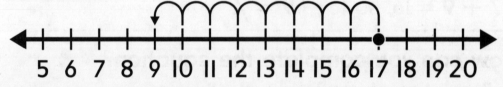

5 6 7 8 9 10 11 12 13 14 15 16 17 18 19 20

Math @ Home Activity

Ask your child to explain the strategy that is best liked when solving
subtraction equations. Write an equation, and have your child teach
you how to use the strategy to solve it.

Additional Practice

Name _____

Review

You know an equation is true when the amounts on each side are equal.

Is this equation true?

$$8 - 3 \overset{?}{=} 10 - 5$$

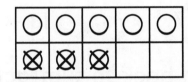

This equation is true.

Is the equation true or false? Circle True or False.

1. $18 - 9 \overset{?}{=} 10$ True   False

2. $11 - 2 \overset{?}{=} 9$ True   False

3. $9 - 7 \overset{?}{=} 10 - 6$ True   False

4. $14 - 9 \overset{?}{=} 12 - 8$ True   False

Complete the equation to make it true.

5. _____ − 4 = 7   **6.** 10 − _____ = 12 − 6

7. 12 − _____ = 4   **8.** 13 − 5 = _____ − 2

9. 17 − 8 = _____   **10.** 14 − 7 = 15 − _____

II. Which equations are true? Circle all of the correct answers..

$12 - 6 \overset{?}{=} 7$ $13 - 10 \overset{?}{=} 9 - 6$

$1 \overset{?}{=} 6 - 5$ $6 \overset{?}{=} 10 - 3$

$15 - 6 \overset{?}{=} 18 - 9$ $10 - 1 \overset{?}{=} 9$

$18 - 8 \overset{?}{=} 9$ $13 - 9 \overset{?}{=} 12 - 7$

Math @ Home Activity

On a sheet of paper, write two columns of true and false subtraction equations. Have your child identify the true and false equations. Then have your child correct the false equations to make them true.

Lesson 6-1
Additional Practice

Name _____

Review

You can describe 2-dimensional shapes by their defining attributes.

Circle the shapes that are closed.

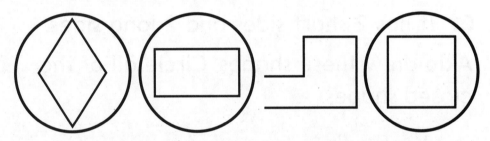

- The second shape is a rectangle. It has 4 vertices and 4 straight sides.

- The last shape is a square. It has 4 vertices. All 4 sides are the same length.

Which shapes match the description? Circle all the correct answers.

1. all sides are same length

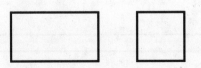

2. shape with 4 vertices

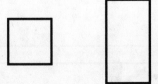

3. Which are *not* true for a rectangle? Choose all of the correct answers.

   A. It has 5 vertices.

   B. It is a closed shape.

   C. It has vertices that are different sizes.

4. Which is *not* true for a square?

   A. It is a 2-dimensional shape.

   B. It has 4 sides.

   C. It has 2 short sides and 2 long sides.

5. Akio drew these shapes. Circle all of the closed shapes.

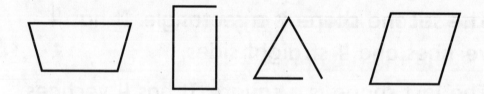

6. Berkley draws a 2-dimensional closed shape with 4 vertices that are the same. Draw a picture to show which shape Berkley could have drawn.

Math @ Home Activity

While outside, point out a shape to your child. Have him or her identify the number of sides and vertices of the specified shape. Repeat with different shapes or objects.

Student Practice Book

68

Additional Practice

Name _____

Review

You can identify 2-dimensional shapes of different colors, sizes, and directions.

Circle the triangles.

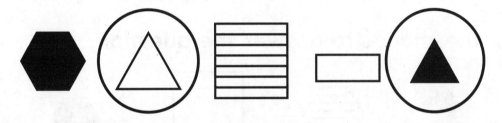

Choose all the correct answers.

I. Circle the hexagons.

2. Circle the circles.

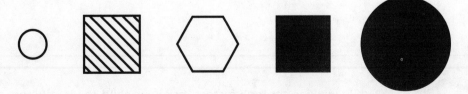

3. Put an X on the squares. Circle the triangles.

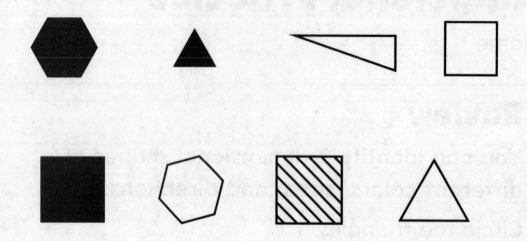

Use the shapes to answer the question.

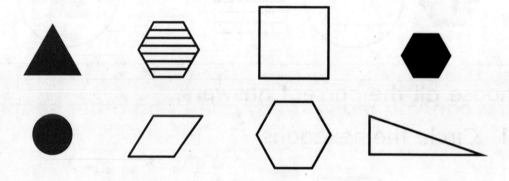

4. How many are hexagons? _____

5. How many have 3 vertices? _____

6. How many have 0 sides? _____

Math @Home Activity

While walking through a store, ask your child to identify two of the same shape that are different colors or sizes, or that are facing different directions. For example, your child may find squares of different sizes and colors as well as squares that are tilted one way or another.

Additional Practice

Name

Review

You can put shapes together to make new shapes.

Make a new shape using these shapes.

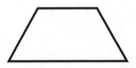  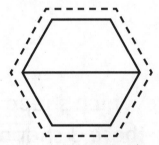

How can you make the shape using other shapes?
Draw lines to show how.

1.

2.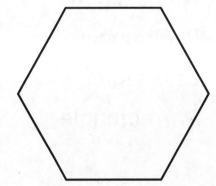

How can you use the shapes to make a new shape? Draw to show how.

3.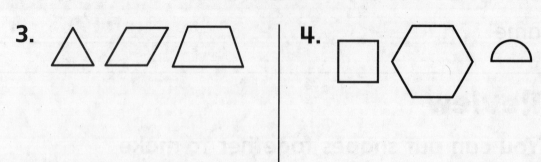

4.

5. Which shape can you make by putting these 3 shapes together?

   A. hexagon

   B. triangle

   C. square

   D. rectangle

Math @ Home Activity

Have your child use shape cutouts to create new shapes using all or a certain number of the shapes. Challenge your child to make different sizes of shapes, such as a larger triangle or a larger square.

Additional Practice

Name _____

Review

You can use the parts of a shape to make a new shape.

Take apart the shape. Then use the parts to make a different shape.

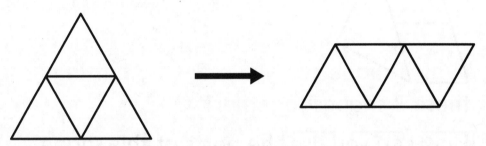

Four triangles were taken apart to make a new shape.

Can you make the pair of shapes from the same parts? Circle Yes or No.

1.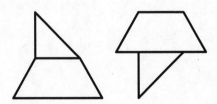

   1   2

   Yes   No

2.

   1   2

   Yes   No

How can you make a different shape using the same parts? Draw to show how.

3.

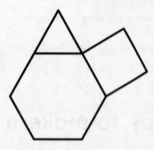

4.

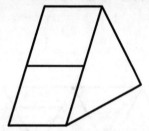

5. How can you use the parts of this shape to make a rectangle? Draw to show your thinking.

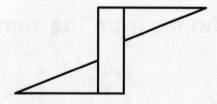

Math @ Home Activity

Provide opportunities for your child to use the parts of a shape to make a different shape. Using the shape cutouts from the last lesson, put some together to make a shape. Then have your child take them apart and use them to create a different shape. Switch roles and repeat.

Additional Practice

Name _____

Review

You can identify a 3-dimensional shape based on the shapes of its faces and the number of faces or bases, edges, and vertices it has.

A cone has 1 apex and 1 circle base.
Circle the cones.

---

1. Circle the cylinders. Put an X on the cubes.

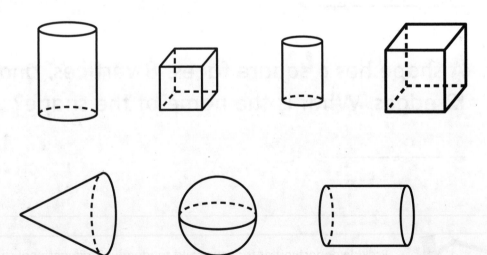

2. Circle the rectangular prisms.

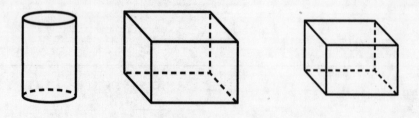

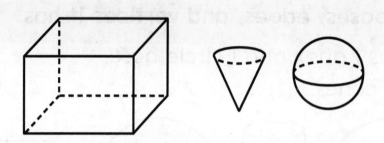

3. A shape has 0 faces, 0 vertices, and 0 edges. What is the name of the shape?

_____

4. A shape has 6 square faces, 8 vertices, and 12 edges. What is the name of the shape?

_____

Additional Practice

Name _____

Review

You can put 3-dimensional shapes together to make new shapes.

Circle the 3-dimensional shapes used to make the larger shape.

 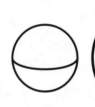  

Circle the shapes that make the larger shape.

1.

2.

 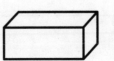

Use the parts of the shape to make a new shape. Draw or describe the new shape.

3.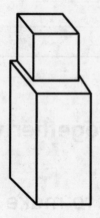

4.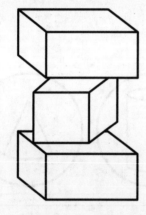

5. Isa wants to make a new shape using the shapes shown. What shape could Isa make?

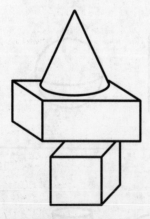

While riding in the car or out on a walk, have your child identify the individual 3-dimensional shapes he or she sees in buildings, statues, and other objects that are composite figures.

Additional Practice

Name _____

Review

You can add one addend to another addend to find the unknown sum.

Harley colors 6 pages. Then he colors 3 more pages.

How many pages does he color?

$$6 + 3 = ?$$

$$6 + 3 = 9$$

He colors 9 pages.

I. Mila catches 8 fish. Then she catches 6 more fish. How many fish does she catch?

Choose all the equations that match the word problem.

A. $6 + 6 = ?$ **B.** $6 + 8 = ?$

C. $8 + 6 = ?$ **D.** $8 + 8 = ?$

2. Rusty swims 9 laps. Then he swims 4 more laps. How many laps does he swim? Draw to show your thinking.

_____ laps

How can you make an equation to solve the problem? Use ? for the unknown. Then solve.

3. 5 children are playing outside. 8 more children join them. How many children are playing outside?

_____ + _____ = _____

_____ children

4. There are 3 birds in a tree. 7 more land in the tree. How many birds are in the tree?

_____ + _____ = _____

_____ birds

5. Dionna sings 6 songs in the morning. She sings 5 songs in the afternoon. How many songs does she sing?

_____ + _____ = _____

_____ songs

Math @ Home Activity

Create addition word problems with your child using scenarios that are familiar to him or her. Have your child write an addition equation using ? to represent the unknown to match each word problem. Have your child use simple drawings to represent and solve the word problems, if needed.

Additional Practice

Name _____

Review

You can use the known addend and a sum to find the unknown addend.

There are 7 books on a desk. A teacher puts some more books on the desk. Now there are 13 books on the desk. How many books did the teacher put on the desk?

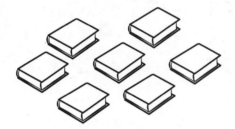 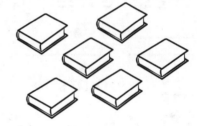

$$7 + ? = 13$$

$$7 + 6 = 13$$

The teacher put 6 books on the desk.

---

1. There are 4 dogs at the park. Some more dogs come to the park. Now there are 12 dogs at the park. How many dogs came to the park? Draw to show your thinking.

_____ dogs

How can you make an equation to show the problem? Use ? for the unknown. Then solve.

2. There are some melons in a wagon. A farmer adds 5 more. Now there are 14 melons in the wagon. How many melons were in the wagon to start?

   _____ + _____ = _____

   _____ melons

3. Emilee needs 12 points to win a game. If she already has 3 points, how many more points does she need to win?

   _____ + _____ = _____

   _____ points

4. Amil sees 6 birds in a tree. He sees some birds on a fence. Amil sees a total of 15 birds. How many birds are on the fence?

   _____ + _____ = _____

   _____ birds

Math @ Home Activity

Have your child gather less than 10 objects, and ask him or her how many more he or she would need to have a certain total number of objects (with the total being 20 or less). For example, if your child has 5 crackers, how many more crackers would he or she need to have 12 crackers? Repeat the activity with different objects and numbers.

Additional Practice

Name _____

Review

You can add two addends, or parts, to find the unknown sum, or whole.

There are 5 small shirts and 9 large shirts. How many shirts are there in all?

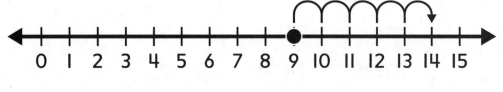

$$5 + 9 = ?$$
$$5 + 9 = 14$$

There are 14 shirts.

Which equation matches the word problem? Circle the equation.

1. There are 7 yellow flowers and 9 blue flowers in a vase. How many flowers are in the vase?

   $$7 + ? = 9 \qquad\qquad 7 + 9 = ?$$

2. Burt collects 5 white eggs and 8 brown eggs. How many eggs does Burt collect?

   $$5 + 8 = ? \qquad\qquad 5 + ? = 8$$

How can you make an equation to show the problem? Use ? for the unknown. Then solve.

3. Clyde has I male duck and 7 female ducks. How many ducks does he have in all?

_____ + _____ = _____

_____ ducks

4. Dasan plants 7 pine trees and 3 oak trees. How many trees does Dasan plant?

_____ + _____ = _____

_____ trees

5. Piper has 8 black cows and 9 brown cows. How many cows does she have?

_____ + _____ = _____

_____ cows

Math @ Home Activity

Give your child many opportunities to solve put together word problems at home. When your child has two variations of the same object, such as red marbles and blue marbles, ask him or her to find the total number of marbles. Repeat with other objects.

Additional Practice

Name _____

Review

You can solve an addition word problem when one or both addends are unknown.

Jana has 14 books. Some are long and the rest are short. How many books are long and how many are short?

Part	Part
?	?
Whole 	

$$14 = ? + ?$$

$$14 = 7 + 7 \qquad\qquad 14 = 10 + 4$$

I. A teacher has 11 pens. She has 5 blue pens. The rest are red pens. How many pens are red? Draw to show your thinking.

_____ red pens

How can you make an equation to show the problem? Use ? for the unknowns. Then solve.

2. There are 15 apples. 8 apples are green and the rest are red. How many red apples are there?

   ____ + ____ = ____

   ____ red apples

3. Silva has blue crayons and brown crayons. She has a total of 12 crayons. How many blue crayons and how many brown crayons does she have?

   ____ = ____ + ____

   ____ blue crayons and ____ brown crayons

4. There are 13 birds. Some are white and some are brown. How many white birds and how many brown birds are there?

   ____ = ____ + ____

   ____ white birds and ____ brown birds

Math @ Home Activity

Present a real-world situation, such as a farmer selling a certain number of vegetables at a market. Ask your child to write an equation to represent the number of each type of vegetable. Then have your child write other equations to represent the situation.

Additional Practice

Name _____

Review

You can add three addends to find an unknown sum.

3 frogs are on a log. 6 frogs are in the water. 2 frogs are on lily pads. How many frogs are there?

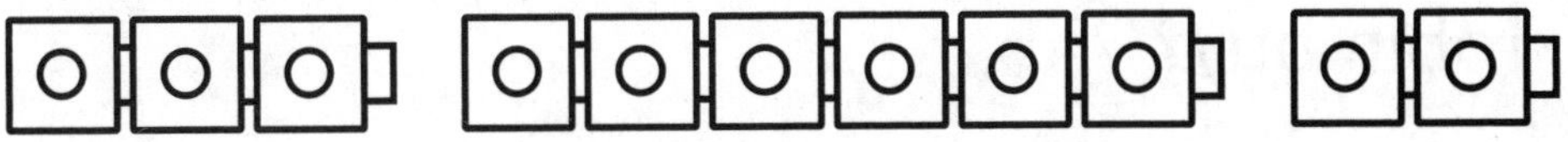

$$3 + 6 + 2 = ?$$
$$3 + 6 + 2 = 11$$

There are 11 frogs.

---

1. Yoko runs 10 miles. She bikes 7 miles. Then she walks 2 miles. How many miles does Yoko travel? Draw to show your thinking.

   _____ miles

How can you make an equation to show the problem? Use ? for the unknown. Then solve.

2. Jayden has 5 animal stickers, 4 car stickers, and 5 sports stickers. How many stickers does Jayden have?

   _____

   _____ stickers

3. Tobin buys 5 peaches, 2 bananas, and 8 oranges. How many pieces of fruit does Tobin buy?

   _____

   _____ pieces of fruit

4. What equation matches the number line? Make an equation.

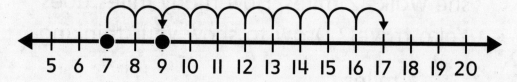

Math @ Home Activity

Provide an addition equation with three addends and ? for the unknown for your child. Have him or her write a word problem to match the equation. Then have your child find the sum to solve the word problem.

Additional Practice

Name _____

Review

You can write an equation with ? for the unknown to solve a word problem.

There are 15 horses. 9 horses were raised on the farm and the rest are new. How many horses are new?

$$15 = 9 + ? \qquad\qquad 15 = 9 + 6$$

6 horses are new.

I. There are 9 kittens. 4 are males and the rest are females. How many female kittens are there?
    How can you make an equation to show the problem? Use ? for the unknown. Then solve.

_____

_____ female kittens

How can you make an equation to show the problem? Use ? for the unknown. Then solve.

2. Hosea picks up 6 seed packets. His dad gives him 7 more packets. How many packets does he have in all?

_____

_____ packets

3. Segio gives away 3 plants. Then he gives away 8 more. Then he gives away 4 more plants. How many plants does he give away in all?

_____

_____ plants

4. Mikaela has 17 berries. Some are blue and 9 are red. How many berries are blue?

_____

_____ blue berries

Look for everyday situations during which you can ask your child to solve addition word problems. For example, if you have 12 envelopes, of which 5 are white and the rest are yellow, ask your child to determine how many envelopes are yellow.

Additional Practice

Name _____

Review

You can subtract one part from the total to find the unknown part, or difference.

There are 13 books on a shelf. 7 books are taken away. How many books are left?

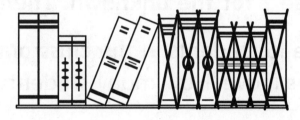

$$13 - 7 = ?$$
$$13 - 7 = 6$$

There are 6 books left.

I. There are 7 fish swimming in a group. 4 fish swim away. How many fish are left? Draw to show your thinking.

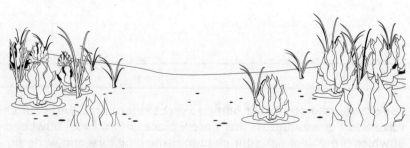

_____ fish

Which equation matches the word problem? Choose all the correct answers.

2. Milos has 14 shells. He gives 6 shells to his sister. How many shells does Milos have now?

   A. $? - 6 = 14$ B. $? = 6 - 14$

   C. $14 - 6 = ?$ D. $? = 14 - 6$

How can you make an equation to show the problem? Use ? for the unknown. Then solve.

3. There are 15 students in the classroom. 9 students leave. How many students are left?

   Equation:

   _____ students

4. There are 17 flowers in a garden. 8 people each pick 1 flower. How many flowers are left?

   Equation:

   _____ flowers

Use situations around your home to create word problems your child can solve. For example, if there are 9 pieces of fruit in a bowl and 2 pieces are eaten, ask your child to draw a picture and write an addition equation to determine how many pieces of fruit are left.

Additional Practice

Name _____

Review

You can use known numbers to find the unknown in a word problem.

There are 12 shirts. Some shirts are taken away. Now there are 8 shirts left. How many shirts were taken away?

$$12 - ? = 8$$
$$12 - 4 = 8$$

4 shirts were taken away.

I. There are 9 deer. Some deer walk away. Now there are 3 deer. How many deer walked away? Draw to show your thinking.

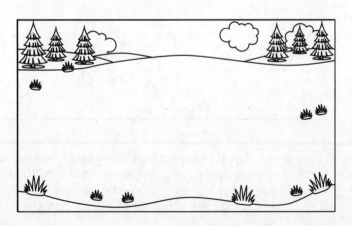

_____ deer

How can you make an equation to show the problem? Use ? for the unknown. Then solve.

2. There are 17 cars in a parking lot. Then some cars left the lot. Now there are 12 cars in the parking lot. How many cars left the lot?

Equation:

_____ cars

3. There are 16 dogs at the park. Then some dogs left. Now there are 8 dogs at the park. How many dogs left the park?

Equation:

_____ dogs

4. There are 14 plates on a table. Some plates are used. Now 9 plates are left. How many plates were used?

Equation:

_____ plates

Math @ Home Activity

Have your child use small objects to solve take from word problems. For example, there are some cups drying in the sink rack and 10 cups are put away. Now 3 cups are in the sink rack. How many cups were in the drying rack to start? Change the situation to match activities around your home.

Additional Practice

Name _____

Review

You can draw to find the unknown total, or whole, in a word problem.

Some shells are on the beach. There are 3 small shells and 14 large shells. How many shells are on the beach?

$$? - 3 = 14$$

$$17 - 3 = 14$$

There are 17 shells on the beach.

---

1. Some stuffed animals are in a bin. There are 3 stuffed cats and 6 stuffed dogs in the bin. How many stuffed animals are in the bin? Draw to show your thinking.

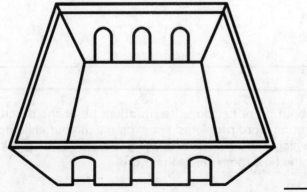

_____ stuffed animals

How can you make an equation to show the word problem? Use ? for the unknown. Then solve.

2. A teacher has some stickers. He has 7 star stickers and 5 balloon stickers. How many stickers does the teacher have?

   Equation:

   _____ stickers

3. Some gifts are on a table. There are 8 blue gifts and 3 pink gifts. How many gifts are there?

   Equation:

   _____ gifts

4. A gardener has some trees. The gardener has 9 pear trees and 7 apple trees. How many trees does the gardener have?

   Equation:

   _____ trees

Math @ Home Activity

Ask your child about his or her day. Use situations he or she describes to create subtraction word problems. For example, if your child has some shirts, of which 7 are white shirts and 8 are black shirts, ask him or her to determine how many shirts there are.

Additional Practice

Name _____

Review

You can solve a word problem when both parts are unknown.

Helga has 9 marbles. Some are silver and some are white. How many are silver and how many are white?

- When both parts are unknown, many equations match the problem.

$$9 - 1 = 8; \quad 9 - 2 = 7; \quad 9 - 3 = 6;$$
$$9 - 4 = 5; \quad 9 - 5 = 4; \quad 9 - 6 = 3;$$
$$9 - 7 = 2; \quad 9 - 8 = 1$$

How can you make an equation to show the problem? Use ? for the unknown.

1. Amee has 18 books. Some are large and some are small. How many are large and how many are small?

   Equation:

2. There are 10 bikes at a store. Some are for children and some are for adults. How many are for children and how many are for adults?

   Equation:

Write three possible answers for the word problem.

3. There are 14 eggs in a chicken coop. Some are white and some are brown. How many are white and how many are brown?

_____ white eggs   _____ brown eggs

_____ white eggs   _____ brown eggs

_____ white eggs   _____ brown eggs

4. There are 17 trees at the park. Some are short and some are tall. How many are short and how many are tall?

_____ short trees   _____ tall trees

_____ short trees   _____ tall trees

_____ short trees   _____ tall trees

Math @ Home Activity

Identify objects around your home that come in two different sizes, colors, or types. Ask your child to identify how many of each type you could have if you have a specific number of both objects combined. You can use red and green apples, two colors of construction paper, small and large baskets, and so on.

Additional Practice

Name _____

Review

You can find an unknown in a word problem.

Luann has 11 books. She has 7 picture books and the rest are chapter books. How many chapter books does Luann have?

Part	Part
●●●● ●●●	**?**
Whole	
●●●●●● ●●●●●●	

$11 - 7 = ?$

$11 - 7 = 4$

Luann has 4 chapter books.

---

1. There are 12 balls in a box. 3 are yellow and the rest are orange. How many orange balls are in the box? Draw to show your thinking.

_____ orange balls

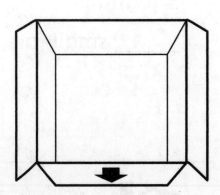

How can you make an equation to show the problem? Use ? for the unknown. Then solve.

2. There are 13 horses in a barn. 8 of the horses are brown and the rest are white. How many horses are white?

Equation:

_____ white horses

3. Roxane has 19 granola bars. 3 have seeds and the rest have berries. How many granola bars have berries?

Equation:

_____ granola bars

4. Dale has 15 jars. 6 are large and the rest are small. How many jars are small?

Equation:

_____ small jars

Math @ Home Activity

When walking outside with your child, look for groups of animals or objects. For example, "There are 10 people at a playground. 4 of them are adults and the rest are children. How many children are at the playground?" Take along a small notebook and a pencil, and have your child write an equation using ? for the unknown. Then have them find the unknown.

Additional Practice

Name _____

Review

You can solve different kinds of subtraction word problems.

There are some swimmers in a pool. 8 of them get out. Now there are 6 swimmers in the pool. How many swimmers were in the pool to start?

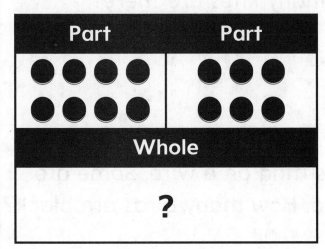

$? - 8 = 6$

$14 - 8 = 6$

There were 14 swimmers in the pool to start.

How can you make an equation to show the problem? Use ? for the unknowns. Then solve.

1. There are 12 carrots. Some are orange and the rest are purple. How many of each color are there?

Equation:

_____ orange carrots and _____ purple carrots

How can you make an equation to show the problem? Use ? for the unknown. Then solve.

2. Lukas has 15 coins. He uses 9 of them to buy a snack. How many coins does he have left?

Equation:

_____ coins

3. There are some kites in the air. 11 of them are higher than the flagpole and 7 are lower than the flagpole. How many kites are there?

Equation:

_____ kites

4. There are 13 birds sitting on a wire. Some are black and 6 are red. How many birds are black?

Equation:

_____ black birds

Math @ Home Activity

Create riddles that match situations in your child's everyday life. Include topics and objects that interest your child. The riddles should require him or her to solve a subtraction problem. For example, "There are 12 frogs in a pond. 3 frogs leave. How many frogs are still in the pond?" Ask your child to draw a representation to solve the riddle.

Additional Practice

Name _____

Review

You can use addition or subtraction to solve word problems.

Havana has 9 jars of paint on a counter. She puts some in a box. Now there are 4 jars of paint on the counter. How many jars did Havana put in the box?

$$9 - ? = 4$$
$$9 - 5 = 4$$

Havana put 5 jars of paint in the box.

How can you make an equation to show the problem? Use ? for the unknowns. Then solve.

I. There are 18 flowers in a vase. Some are pink and the rest are red. How many flowers are pink and how many flowers are red?

Equation:

_____ pink flowers and _____ red flowers

How can you make an equation to show the problem? Use ? for the unknown. Then solve.

2. Vera has 15 spoons. She puts 9 on the counter and some in the drawer. How many spoons does she put in the drawer?

Equation:

_____ spoons

3. There are 19 vehicles in a parking lot. 13 are vans and the rest are trucks. How many are trucks?

Equation:

_____ trucks

4. Some elephants are at a river. 2 elephants walk away. Now there are 11 elephants at the river. How many elephants were at the river to start?

Equation:

_____ elephants

Math @ Home Activity

Create an addition or subtraction word problem for your child to solve. Have him or her show you how to draw a representation to help find the solution to the problem. Ask questions that require your child to demonstrate understanding of his or her problem-solving process.

Additional Practice

Name _____

Review

You can use mental math to help you add 10.

When you add 10, the tens digit goes up by 1 and the ones digit stays the same.

$$23 + 10 = ?$$

$$23 + 10 = \mathbf{33}$$

What is the sum?

1. $10 + 4 =$ _____

2. $71 + 10 =$ _____

3. $29 + 10 =$ _____

4. $10 + 18 =$ _____

5. $44 + 10 =$ _____

6. $10 + 35 =$ _____

Is the equation true? Circle Yes or No.

7. $66 + 10 \stackrel{?}{=} 76$

Yes

No

8. $10 + 7 \stackrel{?}{=} 27$

Yes

No

9. $10 + 59 \stackrel{?}{=} 68$

Yes

No

10. $82 + 10 \stackrel{?}{=} 92$

Yes

No

---

11. There are 39 students in the gym. 10 more students enter the gym. How many students are in the gym now?

_____ students

12. 58 soccer players are practicing on the soccer field. 10 more soccer players begin to practice. How many soccer players are practicing now?

_____ soccer players

Math @ Home Activity

Provide your child with many opportunities to use a pattern to find 10 more than a number. Give your child a number from 1 to 89. Have him or her add 10 to the number and then explain the pattern he or she used. Repeat the activity with a different number.

Additional Practice

Name _____

Review

You can add tens to any number.

When you add tens to a number, you add the tens digits. The ones digit stays the same.

$$14 + 40 = ?$$

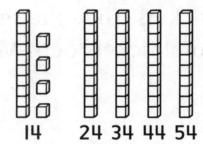

14   24 34 44 54

$$\underline{1}4 + \underline{4}0 = \underline{5}4$$

What is the sum? Fill in the equation.

1. 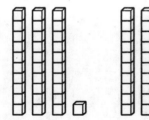 _____ + _____ = _____

2. 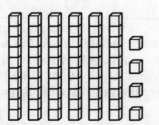 _____ + _____ = _____

What is the sum?

3. $18 + 50 =$ _____

4. $62 + 30 =$ _____

5. $49 + 40 =$ _____

6. $10 + 71 =$ _____

7. $20 + 55 =$ _____

8. $37 + 60 =$ _____

9. Madelenne is selling tickets to the school play. She sells 21 tickets. Then she sells 40 more. How many total tickets does Madelenne sell?

_____ tickets

10. Franz adds $50 + 13$. He says the sum is 53. How can you help Franz find the correct sum?

Math @ Home Activity

Pick a 2-digit number, and have your child write it on a sheet of paper. Then give him and her a certain number of dimes. The number of dimes he or she receives is the number of tens that should be added to the number written on the paper, up to a sum of 99. Have your child add the two numbers. Repeat the activity with different numbers and amounts of dimes.

Additional Practice

Name _____

Review

You can add ones to a number.

When you add ones to a number, you add the ones digits. Since the sum of the ones is less than 10, the tens digit stays the same.

$$31 + 5 = ?$$

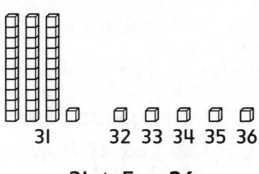

31   32 33 34 35 36

$$3\underline{1} + \underline{5} = 3\underline{6}$$

What is the sum? Fill in the equation.

1. 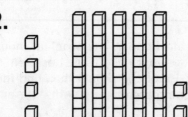   _____ + _____ = _____

2.   _____ + _____ = _____

What is the sum?

3. $57 + 2 =$ _____   4. $4 + 33 =$ _____

5. $21 + 8 =$ _____   6. $62 + 7 =$ _____

7. $5 + 43 =$ _____   8. $52 + 3 =$ _____

9. Soila buys 33 yellow beads and 5 green beads. How many beads does she buy in all?

_____ beads

10. Quinn has 23 trading cards. Her brother gives her 6 more trading cards. How many trading cards does Quinn have in all? Draw to show your thinking.

_____ trading cards

Math @ Home Activity

Write an addition problem on a sheet of paper. The problem should not require your child to regroup. Have your child draw base-ten shorthand to represent the problem. He or she should then find the sum. Repeat with a different addition problem.

Additional Practice

Name _____

Review

You can break apart addends to help add 2-digit numbers.

Add the tens. Add the ones. Then add the tens and ones.

$$24 + 53 = ?$$

$$20 + 50 = 70$$
$$4 + 3 = 7$$
$$70 + 7 = 77$$

24   53

$$24 + 53 = 77$$

What is the sum? Show your thinking.

1.
$$16 + 41 = \underline{\hspace{1.5cm}}$$

2.
$$53 + 25 = \underline{\hspace{1.5cm}}$$

What is the sum? Add the tens. Add the ones. Then add the tens and ones.

3. 36 + 23 = _____

4. 15 + 62 = _____

5. 22 + 26 = _____

6. 45 + 11 = _____

7. Eric exercises for 42 minutes on Monday.
   He exercises for 45 minutes on Tuesday.
   How many minutes does Eric exercise in all?

   _____ minutes

On a sheet of paper, draw base-ten shorthand. Have your child use the base-ten shorthand to solve 2-digit + 2-digit problems. Have your child explain how they found the sum.

Additional Practice

Name _____

Review

You can use an open number line to add numbers.

Show the addition on the open number line. Write the sum.

$$21 + 42 = ?$$

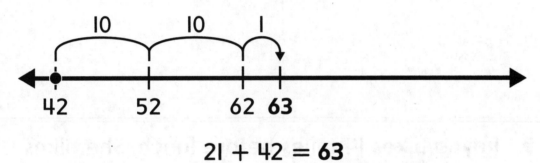

$$21 + 42 = 63$$

What is the sum? Show your thinking on the open number line.

1. $51 + 12 = $ _____

2. $41 + 26 = $ _____

What is the sum?

3. 22 + 57 = _____   4. 13 + 42 = _____

5. 37 + 31 = _____   6. 41 + 54 = _____

7. Rayna bikes 14 miles before lunch. She bikes 25 miles after lunch. How many miles does Rayna bike in all? Show your thinking on the open number line.

_____ miles

Math @ Home Activity

Provide opportunities for your child to use an open number line to add 2-digit numbers. Draw a number line on a dry erase board or sheet of paper. Write a 2-digit + 2-digit equation at the top of the board. Have your child solve the addition problem using the number line. Repeat the activity with different numbers.

Additional Practice

Name _____

Review

You can decompose, or break apart, one addend to make a 10.

$$46 + 9 = ?$$

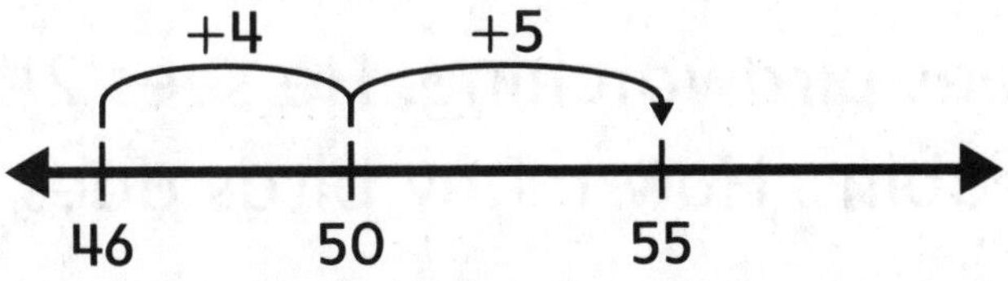

$$46 + 9 = 55$$

What is the sum? Use a number line to show your thinking.

1. $28 + 5 =$ _____

tens	ones

2. $7 + 57 =$ _____

tens	ones

What is the sum?

3. 26 + 6 = _____

4. 65 + 7 = _____

5. 5 + 78 = _____

6. 4 + 39 = _____

7. 42 + 9 = _____

8. 58 + 8 = _____

9. Otto goes birdwatching. He sees 24 sparrows and 8 robins. How many birds does Otto see in all?

_____ birds

10. Suri adds 87 + 6.

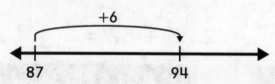

She says the sum is 94. How can you help Suri fix her mistake?

Help your child practice adding with regrouping. Provide your child with an addition problem involving a 1-digit number and a 2-digit number. Have your child draw an open number line to find the sum.

Additional Practice

Name _____

Review

You can regroup to add.

Regroup 10 ones as 1 ten.

Then add the tens and ones.

$$35 + 8 = ?$$

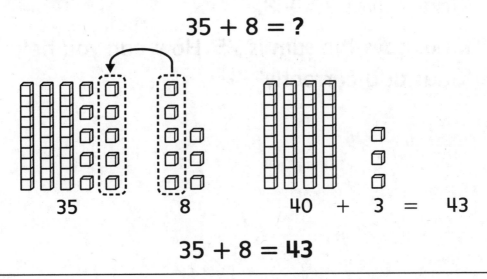

35   8   40 + 3 = 43

$$35 + 8 = 43$$

What is the sum? Show or explain your thinking.

1. $23 + 9 =$ _____

2. $68 + 6 =$ _____

What is the sum?

3. $5 + 87 =$ _____

4. $45 + 6 =$ _____

5. Klaus solves $73 + 8$.

   Klaus says the sum is 75. How can you help Klaus add correctly?

Math @ Home Activity

Have your child solve an addition problem involving a 1-digit and a 2-digit number that requires regrouping, as seen in the exercises on this and the previous page. Encourage your child to make a 10 to find the sum. Ask your child to explain their thinking.

Additional Practice

Name _____

Review

You can regroup to add two 2-digit numbers.

$45 + 16 = ?$

Regroup 10 ones
as 1 ten.

Then add the tens
and ones.

$45 + 16 = 61$

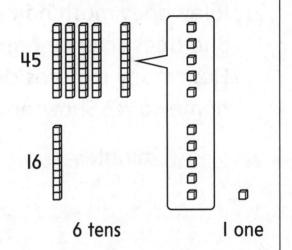

45

16

6 tens   1 one

What is the sum? Draw tens and ones to show your thinking.

1. $28 + 36 =$ _____

2. $14 + 57 =$ _____

What is the sum?

3. $66 + 19 = $ _____

4. $59 + 22 = $ _____

5. $49 + 48 = $ _____

6. $15 + 77 = $ _____

---

7. Riley does math homework for 36 minutes. She does science homework for 46 minutes. How many minutes does Riley spend doing homework? Show or explain your thinking.

_____ minutes

Math @ Home Activity

Write 2-digit plus 2-digit problems on index cards. Have your child sort the cards into two piles: problems that require regrouping to solve and problems that do not. Then have your child solve the addition problems that require regrouping.

Additional Practice

Name _____

Review

You can use addition or subtraction to compare numbers.

Jackson has 11 blocks. Ettie has 5 blocks. How many fewer blocks does Ettie have than Jackson?

Jackson ●●●●●●●●●●●

Ettie ○○○○○

- Solve using subtraction: $11 - 5 = ?$

- Solve using addition: $5 + ? = 11$

Ettie has **6** fewer blocks.

---

1. Fatimah makes 4 wreaths. Alfonso makes 11 wreaths. How many more wreaths does Alfonso make than Fatimah? Draw to show your thinking.

_____ wreaths

How can you make an equation to show the problem? Use ? for the unknown. Then solve.

2. Enid bikes 15 miles. Hassun bikes 7 miles. How many fewer miles does Hassun bike than Enid?

_____ miles

3. Joaquin finds 17 ladybugs. Emery finds 12 ladybugs. How many fewer ladybugs does Emery find than Joaquin?

_____ ladybugs

4. Peta sees 11 ants. Kiego sees 20 ants. How many more ants does Kiego see than Peta?

_____ ants

Math @ Home Activity

Give your child many opportunities to solve compare problems. Create word problems of this type, and ask your child to explain how to solve the problem. Ask him or her to write two equations that can be used to find the answer.

Additional Practice

Name _____

Review

You can add to find the greater number in a word problem.

There are 7 penguins on the ice. There are 3 more penguins in the water than on the ice. How many penguins are in the water?

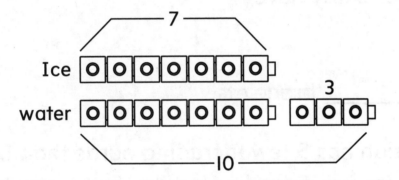

$7 + 3 = ?$

$7 + 3 = 10$

There are 10 penguins in the water.

Make an equation to show the word problem.

1. There are 5 birds in a pine tree. There are 4 more birds in an oak tree than in the pine tree. How many birds are in the oak tree?

Pine Tree

Oak Tree

How can you make an equation to show the problem? Use ? for the unknown. Then solve.

2. Caren has 6 marbles. She has 4 fewer marbles than Gerald. How many marbles does Gerald have?

_____ marbles

3. Chapa has 9 bracelets. She has 6 fewer bracelets than Daisy. How many bracelets does Daisy have?

_____ bracelets

4. Isaiah has 5 fewer trading cards than Devlin. Isaiah has 8 trading cards. How many trading cards does Devlin have?

_____ trading cards

Use the following sentence frame to create addition word problems your child can solve using counters or other small objects. Have your child write an equation for each word problem. <u>Name</u> has <u>number</u> <u>object</u>. <u>Second name</u> has <u>number</u> more <u>object</u>. How many <u>object</u> does <u>second name</u> have?

Additional Practice

Name _____

Review

You can add or subtract to find the lesser number in a word problem.

Zada reads 7 fewer pages than Retta. Retta reads 10 pages. How many pages does Zada read?

Zada ⟨ ? ⟩ ◯◯◯ⓍⓍⓍⓍⓍⓍⓍ

Retta ◯◯◯◯◯◯◯◯◯◯

10

- Solve using subtraction: $10 - 7 = ?$
- Solve using addition: $7 + ? = 10$

Zada reads 3 pages.

Make an equation to show the word problem.

1. Xavier eats 9 more blueberries than Conley. Xavier eats 16 blueberries. How many blueberries does Conley eat?

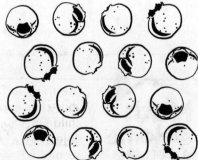

How can you make an equation to show the problem? Use ? for the unknown. Then solve.

2. Eleonore swims 7 fewer laps than Cordell. Cordell swims 14 laps. How many laps does Eleonore swim?

_____ laps

3. Tisha has 10 fewer goldfish than Eliseo. Eliseo has 18 goldfish. How many goldfish does Tisha have?

_____ goldfish

4. Sid sees 3 more bees than Josiah. Sid sees 17 bees. How many bees does Josiah see?

_____ bees

Math @ Home Activity

Create word problems, like the ones above, using situations in your home. Ask your child to solve the word problems using addition or subtraction. Ask for an explanation for the solution of each problem.

Additional Practice

Name _____

Review

You can add or subtract to find different unknowns in a word problem.

Enid has 8 apple slices. Klara has 10 apple slices. How many more apple slices does Klara have than Enid?

- Solve using subtraction: $10 - 8 = ?$
- Solve using addition: $8 + ? = 10$

Klara has 2 more apple slices than Enid.

---

1. Rico works 15 hours. Lu works 6 fewer hours than Rico. How many hours does Lu work? Draw to show your thinking.

_____ hours

How can you make an equation to show the problem? Use ? for the unknown. Then solve.

2. Julie has 18 pencils. Julie has 3 more pencils than Dora. How many pencils does Dora have?

_____ pencils

3. George swims 5 laps in the pool. Declan swims 13 laps. How many fewer laps does George swim than Declan?

_____ laps

4. Waldo has 9 fewer toy dinosaurs than Uma. Uma has 18 toy dinosaurs. How many toy dinosaurs does Waldo have?

_____ toy dinosaurs

Math @ Home Activity

Look for comparison situations in everyday life that can be turned into a word problem that your child can solve. Encourage your child to draw a picture and write at least one equation to solve the problem.

Additional Practice

Name _____

Review

You can subtract 10 from a 2-digit number.

- Use a number chart to find 52 − 10.

- The tens digit goes down by 1.

- The ones digit stays the same.

52 − 10 = 42

1	2	3	4	5	6	7	8	9	10
11	12	13	14	15	16	17	18	19	20
21	22	23	24	25	26	27	28	29	30
31	32	33	34	35	36	37	38	39	40
41	**42**	43	44	45	46	47	48	49	50
51	52	53	54	55	56	57	58	59	60
61	62	63	64	65	66	67	68	69	70
71	72	73	74	75	76	77	78	79	80
81	82	83	84	85	86	87	88	89	90
91	92	93	94	95	96	97	98	99	100

What is the difference?

1. 23 − 10 = ? _____

2. 87 − 10 = ? _____

3. 99 − 10 = ? _____

4. 45 − 10 = ? _____

5. 54 − 10 = ? _____

6. 76 − 10 = ? _____

1	2	3	4	5	6	7	8	9	10
11	12	13	14	15	16	17	18	19	20
21	22	23	24	25	26	27	28	29	30
31	32	33	34	35	36	37	38	39	40
41	42	43	44	45	46	47	48	49	50
51	52	53	54	55	56	57	58	59	60
61	62	63	64	65	66	67	68	69	70
71	72	73	74	75	76	77	78	79	80
81	82	83	84	85	86	87	88	89	90
91	92	93	94	95	96	97	98	99	100

7. Santo has 31 marbles. He gives his sister 10 marbles. How many marbles does Santo have left?

_____ marbles

8. There are 48 people at the zoo. 10 people leave. How many people are at the zoo now?

_____ people

9. Malika makes 62 care packages. She gives 10 away. How many care packages does she have left?

_____ care packages

10. Emil jogs for 36 minutes. Willa jogs 10 fewer minutes than Emil. Willa says she jogged for 27 minutes. How can you help Willa subtract 36 − 10?

Additional Practice

Name _____

Review

You can subtract tens.

50 − 40 = ?

Subtract 40 by taking away 4 tens.

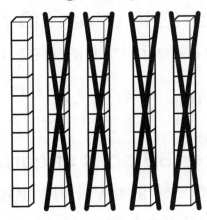

$$\underline{5}0 - \underline{4}0 = 10$$

What subtraction equation matches the tens blocks? Write the equation.

1.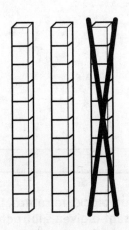

   _____

2.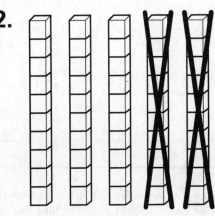

   _____

What is the difference? Show your thinking.

3. 90 − 30 = _____

4. 70 − 50 = _____

5. Eloy sees 40 birds in a tree. 30 birds fly away. How many birds are in the tree now?

_____ birds

6. Madalyn makes 80 pairs of earrings. She sells 60 pairs at a craft show. How many pairs of earrings does she have left?

_____ pairs of earrings

Ask your child to explain how to use place-value blocks to subtract tens. Have him or her use the cutouts to represent a given subtraction problem. Have your child write a subtraction equation to match the subtraction performed. Repeat with a different subtraction problem.

Additional Practice

Name _____

Review

You can use a number chart or open number line to subtract tens.

Use a number chart to find 50 − 30.

1	2	3	4	5	6	7	8	9	10
11	12	13	14	15	16	17	18	19	**20**
21	22	23	24	25	26	27	28	29	30
31	32	33	34	35	36	37	38	39	40
41	42	43	44	45	46	47	48	49	**50**
51	52	53	54	55	56	57	58	59	60
61	62	63	64	65	66	67	68	69	70
71	72	73	74	75	76	77	78	79	80
81	82	83	84	85	86	87	88	89	90
91	92	93	94	95	96	97	98	99	100

50 − 30 = ?

50 − 30 = **20**

What is the difference?

1. 80 − 40 = ? _____

2. 60 − 30 = ? _____

3. 70 − 50 = ? _____

4. 90 − 80 = ? _____

5. 40 − 20 = ? _____

6. 30 − 30 = ? _____

7. 50 − 40 = ? _____

1	2	3	4	5	6	7	8	9	10
11	12	13	14	15	16	17	18	19	20
21	22	23	24	25	26	27	28	29	30
31	32	33	34	35	36	37	38	39	40
41	42	43	44	45	46	47	48	49	50
51	52	53	54	55	56	57	58	59	60
61	62	63	64	65	66	67	68	69	70
71	72	73	74	75	76	77	78	79	80
81	82	83	84	85	86	87	88	89	90
91	92	93	94	95	96	97	98	99	100

Use a number line to subtract tens.
Write the difference.

8. 80 − 50 = ? _____

9. 90 − 70 = ? _____

10. 40 − 40 = ? _____

11. Booker prints 70 schedules for an event. He gives out 60 schedules. How many schedules does he have left?

_____ schedules

Math @ Home Activity

To gain practice subtracting tens, have your child explain how to use an open number line to find a difference. Have him or her solve a subtract tens from tens problem, explaining each step along the way. Then have your child provide a subtraction problem for you to solve on an open number line, using the same steps he or she explained.

Additional Practice

Name _____

Review

You can use addition and known facts to help you subtract tens.

To solve $80 - 50 = ?$, you can make an unknown addend equation: $50 + ? = 80$.

Then use a known fact to solve: $5 + 3 = 8$.

I know $50 + \mathbf{30} = 80$, so $80 - 50 = \mathbf{30}$.

How can you use addition to find the difference? Show your thinking.

1. $20 - 20 =$ _____

2. $60 - 30 =$ _____

3. $40 - 10 =$ _____

4. $70 - 60 =$ _____

What addition equation can you use to help you subtract? Write the equation.

5. $90 - 10 = ?$

6. $50 - 40 = ?$

7. $60 - 60 = ?$

8. $80 - 20 = ?$

9. Woodrow makes 40 cat toys. He donates 40 cat toys to an animal shelter. How many cat toys does Woodrow have left?

_____ cat toys

10. Earl picks 30 carrots. His dad uses 20 carrots to make soup. How many carrots are left?

_____ carrots

Additional Practice

Name

Review

You can use different strategies to subtract tens.

There are 70 people on a bus. 30 people get off the bus. How many people are on the bus now?

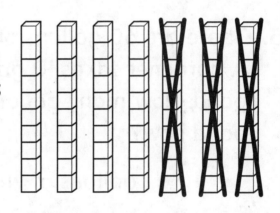

$\underline{7}$ tens $-$ $\underline{3}$ tens $=$ $\underline{4}$ tens

$70 - 30 = \mathbf{40}$

There are 40 people on the bus now.

Is the equation true? Circle Yes or No.
Explain how you know.

I. $80 - 50 \overset{?}{=} 30$

    Yes   No

2. $20 - 20 \overset{?}{=} 10$

    Yes   No

What is the difference? Explain or show your thinking.

3. 50 − 20 = _____

4. 90 − 40 = _____

5. There are 60 gallons of water in a garden pond. A gardener takes 10 gallons of water from the pond. How many gallons of water are in the pond now?

_____ gallons of water

Additional Practice

Name _____

Review

You can compare the lengths of objects.

Circle the longest pencil.

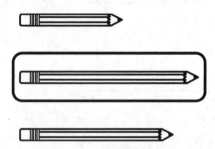

Make sure the pencils' ends are lined up to compare the lengths.

---

1. Circle the longer paintbrush.

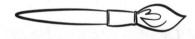

2. Alton and Myron are using markers to draw. Alton is using a shorter marker than Myron. Circle Alton's marker.

How can you order the objects by length?
Write A, B, or C.

A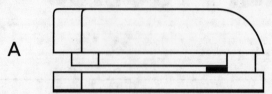

B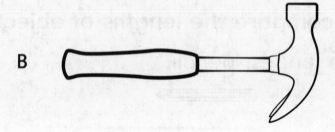

C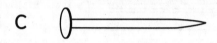

3. Which object is shortest?

Object _____ is shortest.

4. Which object is longest?

Object _____ is longest.

5. How can you compare the lengths of the objects?

Object A is shorter than object _____.

Object A is longer than object _____.

Math @ Home Activity

Align the endpoints of kitchen items, such as a fork, spoon, and spatula. Have your child compare the items' lengths. Ask your child to identify the shortest and longest items and explain their reasoning.

Additional Practice

Name

Review

You can compare the lengths of two objects by using a third object.

Is the crayon longer than or shorter than the marker?

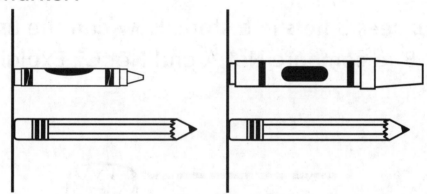

The pencil is longer than the crayon but shorter than the marker. This means that the crayon is shorter than the marker.

---

1. Is the ladder longer than or shorter than the shovel?

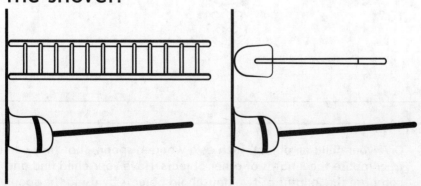

The ladder is _____ than the shovel.

2. Is the football longer than or shorter than the megaphone?

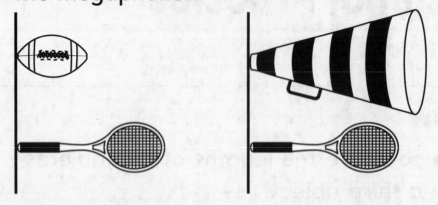

The football is _____ than the megaphone.

3. Fleur sees 3 nets in a store. How can she use Net B to compare Net A and Net C? Explain.

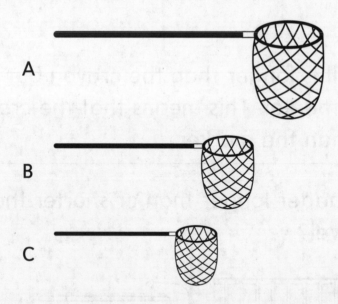

A

B

C

Math @ Home Activity

Give your child an object, such as a wooden spoon, that they can use to compare the lengths of other objects. Have your child find and compare the lengths of two household objects by using the spoon as a third object. Repeat with different household objects.

Student Practice Book
142

Additional Practice

Name _____

Review

You can use units to measure length.

- Measure the length of the whistle using paper clips.

- Lay the paper clips end to end with no gaps or overlaps.

The whistle is 2 paper clips long.

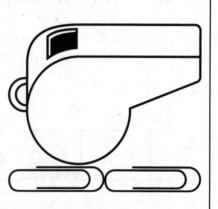

---

1. How many pennies long is the marker?

_____ pennies

2. How many ones units long is the key?

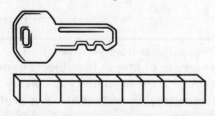

_____ ones units

3. How long is the spoon?

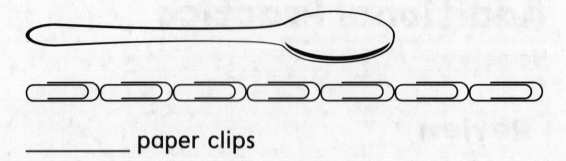

_____ paper clips

4. Uri says the stapler is 9 nickels long.

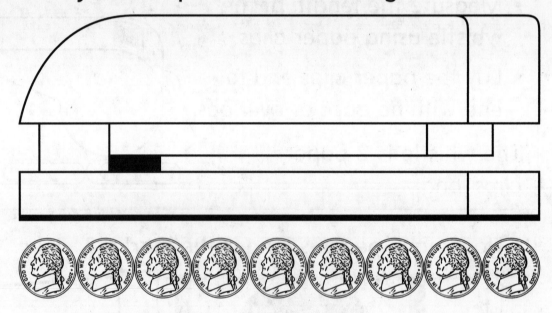

Do you agree with Uri? Explain.

Math @ Home Activity

Have your child gather toys or other objects from your home to measure. Incorrectly measure the length of one of the objects using same-size units such as nickels or blocks. For example, you can lay the units along the length of the object with gaps or overlaps. Ask your child to correct your mistake and then correctly measure that object and the other objects they collected.

Additional Practice

Name _____

Review

You can use different units to measure the same object.

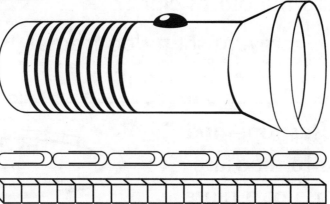

The flashlight is 6 paper clips long.

The flashlight is 18 ones units long.

---

1. How long is the pot?

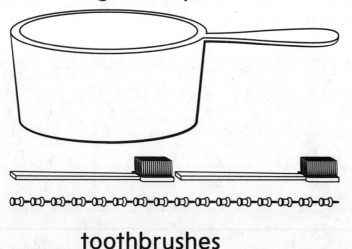

_____ toothbrushes

_____ push pins

How can you use two different units to measure the alarm clock?

2a. Which unit is larger?

    A. dog treats   **B.** crayons

2b. Will it take *more* or *fewer* crayons than dog treats to measure the alarm clock?

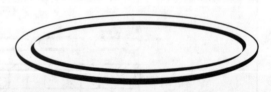

_____ crayons than dog treats

3. Dinah uses tape and peanuts to measure the length of a plate. Will she use *more* or *fewer* peanuts than tape to measure the plate? Explain.

_____ peanuts than tape

Additional Practice

Name ....

Review

You can use clocks to tell time to the hour.

What time is it on the analog clock?

- The minute hand points to 12.

- The hour hand points to 9.

The clock shows 9 o'clock.

What time is it? Write the time.

1.

_____ : _____

2.

_____ : _____

3. Leona looks at the clock at the beginning of tennis practice. What time is showing?

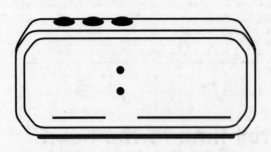

___

4. Kendric's family has dinner at 6 o'clock. What time does dinner start?

_____:_____

5. The movie starts at I o'clock. What time does it start?

_____:_____

Additional Practice

Name _____

Review

You can use clocks to tell time to the half hour.

What time is shown on the clock?

- The minute hand points to 6.
- The hour hand is between 2 and 3.

The time is 2:30, or half past 2.

What time is it? Write the time.

1.

____:____

2.

____:____

3. Mr. Fleming looks at the clock when he puts muffins in the oven to bake. What time is showing?

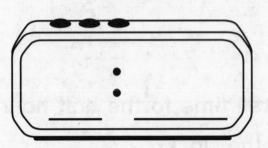

4. Swim practice starts at half past 10:00. What time does it start?

_____:_____

5. Darby says the time is half past 12:00. Pablo says the time is 12:30. How can they both be correct?

Additional Practice

Name _____

Review

You can sort objects into categories in a chart.

There are different ways to sort objects. These objects were sorted by shape.

Triangle	Rectangle	Circle

---

1. How can you sort these objects? Complete the chart. Explain how you organized the data.

sock   yo-yo   drum   teddy bear   trumpet   shirt

Clothing	Toy	Instrument

Draw the objects or write their names to complete the chart.

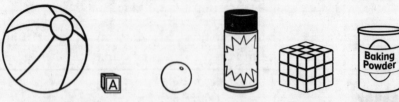

beach ball   block   orange   spray can   puzzle cube   baking powder

2a. How can you sort the objects by shape?

Cylinder	Sphere	Cube

2b. How can you sort the objects by size?

Small	Medium	Large

Math @ Home Activity

Gather a group of household objects. Ask your child to sort them in one way, such as by size, shape, or use. Challenge your child to sort the objects in at least one other way. Then have your child gather objects for you to sort in different ways. Sort the objects, and then have your child guess how you sorted them.

Additional Practice

Name _____

Review

You can show how many of each kind of object in a tally chart.

How many people were asked to name their favorite zoo animal?

Favorite Zoo Animal		
Animal	**Tally**	**Total**
	卌 ǀǀ	7
	卌	5
	卌 ǀ	6

7 people like turtles.

5 people like chimps.

6 people like giraffes.

$$7 + 5 + 6 = 18$$

18 people were asked to name their favorite zoo animal.

1a. How many people chose each type of fruit as their favorite? Write the totals.

1b. How many people chose a favorite fruit?

_____ people

Favorite Fruit		
Fruit	**Tally**	**Total**
	卌 ǀǀǀǀ	
	卌 ǀ	
	ǀǀǀ	

Some students chose their favorite type of weather.

2a. How can you show the data in a tally chart? Make tally marks and write numbers to complete the chart.

Favorite Weather		
Weather	**Tally**	**Total**
☀		
🌧		
❄		

2b. How many students chose their favorite type of weather?

_____ students

Math @ Home Activity

Have your child ask family and friends a question that will produce data that can be displayed in a chart. For example, your child might ask people to name their favorite subject or favorite season. Have him or her create a chart to display the answers. Ask your child how many people he or she surveyed to create the chart.

Additional Practice

Name _____

Review

You can use a tally chart to answer questions about data.

Shen made a tally chart about his friends' favorite sports. How many like volleyball the best?

Sport		Tally
🏐	Baseball	||||
⚽	Soccer	|||
🏐	Volleyball	⊬|

There are 6 tallies. So, 6 like volleyball the best.

I. The swim team voted for their favorite type of swim stroke. How many people voted?

_____ people

Swim Stroke		Tally
	Backstroke	⊬|
	Freestyle	⊬||
	Sidestroke	||||

2. Zuri makes a tally chart to show her classmates' favorite flowers. Use the tally chart to answer the questions.

Flower	Tally
Rose 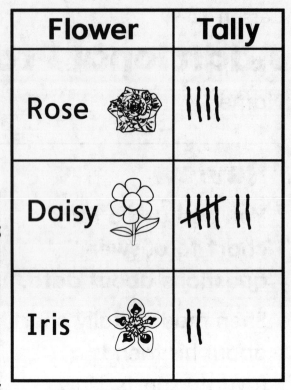	IIII
Daisy	IIII II
Iris	II

a. How many students like daisies the best?

_____ students

b. How many students like roses the best?

_____ students

c. How many students like irises the best?

_____ students

d. How many students voted in all?

_____ students

Math @ Home Activity

Use pennies, nickels, and dimes to create a tally chart to display how many of each type of coin. Have your child come up with questions that can be answered using the data.

Additional Practice

Name _____

Review

You can use a tally chart to answer questions about data.

How many children voted?

- 5 voted for airplane.

- 4 voted for doll.

- 6 voted for jump rope.

$5 + 4 + 6 = 15$, so 15 children voted.

Toy	Tally
✈	卌
👧	卌
➰	卌 l

---

1. Use the tally chart to answer the questions.

   a. Which insect did the fewest choose? _____

   b. Which insect did the most choose? _____

   c. How many more students chose butterfly than ant? _____ students

Insect		Tally
Ant	🐜	卌 l
Bee	🐝	卌 ll
Butterfly	🦋	卌 lll

2. Use the tally chart to answer the questions.

Tools	Tally
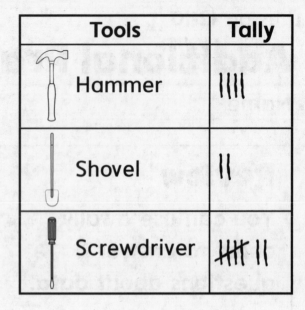 Hammer	IIII
Shovel	II
Screwdriver	IIII II

a. How many fewer students chose shovel than screwdriver?

_____ students

b. How many more students chose hammer than shovel?

_____ students

c. How many students chose a favorite tool?

_____ students

Math @ Home Activity

With your child, create a tally chart displaying how many of three different types of animals you see within one day. Ask your child which animal was seen the most and which was seen the least. Then ask them how many more times one animal was seen than another.

Additional Practice

Name _____

Review

You can tell whether parts of a shape are equal.

Does the shape have equal shares?
Circle Yes or No.

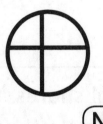   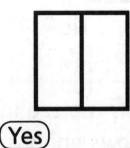

Yes   (No) | (Yes)   No

Does the shape have equal shares?
Circle Yes or No.

1.

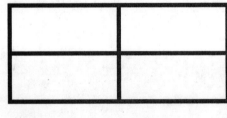

Yes   No

2.

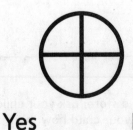

Yes   No

3.

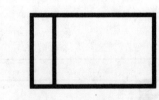

Yes   No

Does the shape have equal shares?
Circle Yes or No.

4.

Yes   No

5.

Yes   No

---

6. Betty cuts a pizza into equal shares. Which pizza shows equal shares? Circle the correct pizza.

Math @ Home Activity

While walking through the park or through a store, ask your child to identify shapes that have equal shares. Ask your child how many equal shares the shape has. Point out shapes that do not have equal shares and ask your child to explain why the shares are not equal.

Additional Practice

Name _____

Review

You can draw a line to make halves.

A whole with 2 equal shares shows halves.

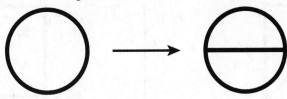

Which shape shows halves? Circle the shape.

1.

2.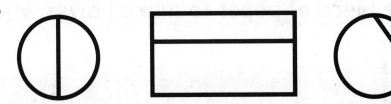

3.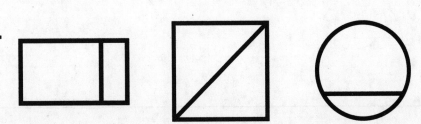

How can you make halves? Draw to show halves.

4.

5.

Does the shape show halves? Circle Yes or No.

6.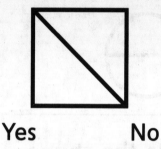

Yes   No

7.

Yes   No

8. Steve wants to cut a piece of paper into halves.
Draw a line on the rectangle to show how Steve
can cut the piece of paper to make halves.

Math @Home Activity

Look for situations around your home where your child can identify or create halves. For example, have your child use their finger to show how objects around your home can be divided in half. Ask them how many equal shares the object is divided into when a shape is divided into halves.

Additional Practice

Name _____

Review

You can draw lines to make fourths.

A whole with 4 equal shares shows fourths, or quarters.

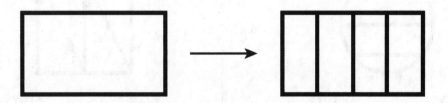

Which shape shows quarters? Circle the shape.

1.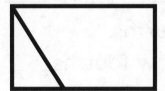

2.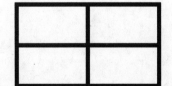

3.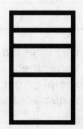

How can you make fourths? Draw to show fourths.

4.

5.

Does the shape shows fourths? Circle Yes or No.

6.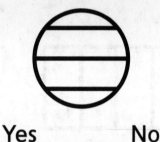

Yes   No

7.

Yes   No

8. Carmen divides a circle into fourths.
Draw on the circle below to show fourths.

Math @ Home Activity

Prepare paper cutouts of squares, rectangles, and circles. Have your child experiment with different ways to fold the cutouts into fourths. For example, your child will find that there are at least three ways to fold a square into fourths.

Additional Practice

Name _____

Review

You can count and describe the equal shares in a whole.

How many quarters are there?

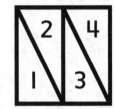

There are 4 quarters, or 4 fourths.

---

1. Circle the shape that has 4 fourths.

2. Circle the shape that has 2 halves.

How many equal shares are there?
Write the number.

3.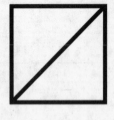

_____ halves

4.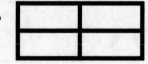

_____ quarters

How can you describe the number of equal shares in the whole?

5.

_____

6.

_____

7. Keegan says there are 2 equal shares.
How can you help Keegan fix his mistake?

_____

 Math @ Home Activity

On a dry erase board or a sheet of paper, draw shapes divided into halves and fourths, as well as shapes that are divided into 2 or 4 unequal shares. Have your child identify how many equal shares are in each whole, whether it is 0 equal shares, 2 halves, or 4 fourths.

Additional Practice

Name _____

Review

You can show halves and fourths of the same whole.

- When a whole has fewer equal shares, the shares are larger. Halves are larger than fourths.

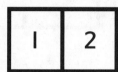

- When a whole has more equal shares, the shares are smaller. Fourths are smaller than halves.

I. Which shows larger equal shares?

2. Which shows smaller equal shares?

 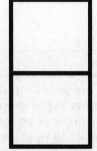

3. Which shows fewer equal shares?

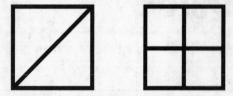

4. Which shows more equal shares?

5. Which sandwich shows fewer equal shares? Circle the sandwich. Complete the sentence.

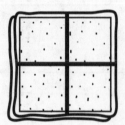 

When a whole has fewer equal shares, the shares are _____.

Cut two rectangles, two squares, and two circles out of paper. Ask your child to draw to divide one set of the shapes into halves. Repeat the activity by dividing the other set of shapes into fourths. Have your child compare the size and number of equal shares in each set of shapes. For example, ask your child to identify which square has smaller equal shares and whether that means it has fewer or more equal shares.